Advances in Intelligent Systems and Computing

Volume 422

Series editor

Janusz Kacprzyk, Polish Academy of Sciences, Warsaw, Poland
e-mail: kacprzyk@ibspan.waw.pl

About this Series

The series "Advances in Intelligent Systems and Computing" contains publications on theory, applications, and design methods of Intelligent Systems and Intelligent Computing. Virtually all disciplines such as engineering, natural sciences, computer and information science, ICT, economics, business, e-commerce, environment, healthcare, life science are covered. The list of topics spans all the areas of modern intelligent systems and computing.

The publications within "Advances in Intelligent Systems and Computing" are primarily textbooks and proceedings of important conferences, symposia and congresses. They cover significant recent developments in the field, both of a foundational and applicable character. An important characteristic feature of the series is the short publication time and world-wide distribution. This permits a rapid and broad dissemination of research results.

More information about this series at http://www.springer.com/series/11156

Paolo Ciancarini · Alberto Sillitti
Giancarlo Succi · Angelo Messina
Editors

Proceedings of 4th International Conference in Software Engineering for Defence Applications

SEDA 2015

 Springer

Editors
Paolo Ciancarini
Dipartimento di Informatica
University of Bologna
Bologna
Italy

Alberto Sillitti
Center for Applied Software Engineering
and CINI
Bolzano
Italy

Giancarlo Succi
Innopolis University
Kazan, Tatarstan Republic
Russia

Angelo Messina
Logistic Department of Italian Army
Rome
Italy

ISSN 2194-5357   ISSN 2194-5365   (electronic)
Advances in Intelligent Systems and Computing
ISBN 978-3-319-27894-0   ISBN 978-3-319-27896-4   (eBook)
DOI 10.1007/978-3-319-27896-4

Library of Congress Control Number: 2015958323

Preface

The military world has always shown great interest in the evolution of software and in the way it has been produced through the years. The first standard for software quality was originated by the US DOD (2167A and 498) to demonstrate the need for this particular user to implement repeatable and controllable processes to produce software to be used in high-reliability applications.

Military systems rely more and more on software than older systems did. For example, the percentage of avionics specification requirements involving software control has risen from approximately 8 % of the F-4 in 1960 to 45 % of the F-16 in 1982, 80 % of the F-22 in 2000, and 90 % of the F-35 in 2006. This reliance on software and its reliability is now the most important aspect of military systems. The area of application includes mission data systems, radars/sensors, flight/engine controls, communications, mission planning/execution, weapons deployment, test infrastructure, program lifecycle management systems, software integration laboratories, battle laboratories, and centers of excellence. Even if it is slightly less significant, the same scenario applies to the land component of the armed forces. Software is now embedded in all the platforms used in operations, starting from the wearable computers of the dismounted soldier up to various levels of command and control, and every detail of modern operations relies on the correct behavior of some software product.

Many of the mentioned criticalities are shared with other public security sectors such as the police, the firefighters, and the public health system. The rising awareness of the critical aspects of the described software diffusion convinced the Italian Army General Staff that a moment of reflection and discussion was needed and with the help of the universities, the SEDA conference cycle was started.

For the third conference SEDA 2014, it was decided to shift the focus of the event slightly away from the traditional approach to look at innovative software engineering. Considering the title: software engineering for defense application, this time, the emphasis was deliberately put on the "defense application" part. For the first time, papers not strictly connected to the "pure" concept of software engineering, were accepted together with others that went deep into the heart of this science.

The reasons for this change were first of all the need for this event to evolve and widen its horizon and secondly the need to find more opportunities for the evolution of military capabilities. In a moment of economic difficulty, it is of paramount importance to find new ways to acquire capabilities at a lower level of funding using innovation as a facilitator. It was deemed very important, in a period of scarce resources to look ahead and leverage from dual use and commercial technologies.

Software is, as said, a very pervasive entity and is almost everywhere, even in those areas where it is not explicitly quoted. A mention was made to the changes in the area of software engineering experienced in the Italian Army and the starting of a new methodology which would then become "Italian Army Agile."

The base of the new approach was presented this way by Gen. Angelo Messina in his introductory speech:

> In the commercial world, "Agile" software production methods have emerged as the industry's preferred choice for innovative software manufacturing. All the Android apps and similar software are generated using short and lean production cycles. Lately Agile practices seem to be in line with the objectives the USA DoD is trying to achieve with the reforms directed by Congress and DoD Acquisition Executives.
>
> DoD Instruction 5000.02 (Dec 2013) heavily emphasizes tailoring program structures and acquisition processes to the program characteristics. At the same time, in May 2013, the Italian Army started looking for a way of solving the problem of the volatility of the user requirement that is at the base of the software development process. The area considered was the Command and Control one where a good deal of lessons learned were available from operations in Iraq and Afghanistan. It was observed that the mission needs in the area were not only changing from one mission to another but also in the time frame of the same mission.
>
> It was evident that the traditional "waterfall" software factories approach was not usable any more. Agile development methods seemed to be capable of deploying quicker and less risky production lines by the adoption of simple principles:
>
> - Responding rapidly to changes in operations, technology, and budgets;
> - Actively involving users throughout development to ensure high operational value;
> - Focusing on small, frequent capability releases;
> - Valuing working software over comprehensive documentation.
>
> Agile practices such as SCRUM include: planning, design, development, and testing into an iterative production cycle (Sprint) able to deliver working software at short intervals (3 weeks). The development teams can deliver interim capabilities (at demo level) to users and stakeholders monthly. These fast iterations with user community give a tangible and effective measure of product progress meanwhile reducing technical and programmatic risk. Response to feedback and changes stimulated by users is far quicker than using traditional methods.
>
> The User/stakeholder community in the Army is very articulated, including Operational Units, Main Area Commands, and Schools. The first step we had to take was the establishment of a governance body which could effectively and univocally define the "Mission Threads" from which the support functions are derived.

Our first Scrum team (including members from Industry) was established in March 2014 and until now has successfully run 8 production cycles. Let me say proudly that there are not many similar production lines in the military software arena. Of course the introduction of the Agile Scrum methodology was not easy nor simple to be worked out. A relevant cultural change is required to switch from a Program Management, time goal, approach to a team centric, customer satisfaction, approach. I cannot say today that the process is concluded but we have enough confidence in the method now to see a clear way ahead.

We are ready to share, with our industrial partners the results of our experience to help them build solid and permanent agile production teams.

Contents

Managing Increasing User Needs Complexity Within the ITA Army Agile Framework

Franco Raffaele Cotugno

Abstract The innovations introduced in the Italian Army in Software (SW) Development Methodologies in the context of the "ITA ARMY AGILE" (IAA) initiative require attentive and tailored governance with regard to the growth of the enterprise complexity and the future requirements related to the development of Command and Control systems. Starting from a limited approach centered on a single Integrated Development Team (IDT), an unprecedented growth of the product is being experienced due essentially to the fact that the customer has been "educated" and is now capable of stating its needs. Consequently, in the positive environment created and witnessed by the change of mentality, a brief description of the issues, chiefly but not solely linked to the scarcity of resources, time management, and procedures complexity, will be provided in order to take the whole Land Command and Control Evolution (LC2EVO) under the expected control.

1 Introduction

Starting from 2014, a new methodology, strongly supported by the Army General Staff (AGS) Logistic Department (LD), has unquestionably served to change the Italian Army's way of procuring/developing software. Due mainly, but not solely to the fact that military requirements are hardly prone to be defined too much in advance but rather are to be scoped in a repetitive cycle of continuous refinements, the Italian Army has approached and ultimately implemented an Agile Software Development Methodology (AM) [1, 2].

Franco Raffaele Cotugno—Italian Army General Staff (IAGS).

F.R. Cotugno (✉)
Stato Maggiore dell'Esercito Italiano (SME), Rome, Italy
e-mail: franco.cotugno@esercito.difesa.it

© Springer International Publishing Switzerland 2016
P. Ciancarini et al. (eds.), *Proceedings of 4th International Conference in Software Engineering for Defence Applications*, Advances in Intelligent Systems and Computing 422, DOI 10.1007/978-3-319-27896-4_1

Among a number of AMs, SCRUM has proved to better fit the specific environment leveraging onto its inherent ability of successfully managing and accommodating ever-changing requirements [3].

The initial implementation has pointed out some incompatibilities between the traditional SCRUM and the specificities of the Army environment to be essentially linked back to the bidimensional nature of the organization involving both military and civilians as well as to the complexity of the defense structure involving multiple threads. As a direct consequence of these constrains, ITA ARMY AGILE (IAA) has been implemented in the form of a customized approach to the SCRUM methodology specifically tailored to address the Army needs [4].

Therefore, IAA has been firstly considered as an opportunity to be investigated and only after a positive assessment with regard to its feasibility to address Army requirements had been made, it was gradually implemented, hence providing for an opportunity to demonstrate its ability in timely delivering quality software focusing on value from the user's perspective [5–7]. The experience, thus far, has provided evidence of the fact that the IAA applied to military systems whose reliability constitutes an element of paramount importance is also instrumental in controlling software development costs and length of the development cycles, and ultimately generates customers' satisfaction [7, 8].

Last year's activities have mainly focused on reinforcing experiences, approaching the cultural change beneficial to all of community of interest (COI), and persuading and "educating" people with respect to the need to remain at the center of the process, as both the requirements' owner and the final user. Having matured a clear understanding of the process and acquired an unambiguous idea of their needs enabled people to support the methodology.

Consequently, the mission of the AGS LD has been of significant complexity in, concurrently, educating people, delivering incremental "pieces" of capabilities all the while monitoring, assessing, and eventually refining processes with the aim to enhance their effectiveness.

The stepwise refinement process has been put in place, and it has proved that within the Italian Army, IAA can work effectively only if it is capable of managing its growth [9].

All these above-mentioned aspects will be explored in this paper, including the whole process of managing the IAA to develop the new Command and Control System (LC2EVO—Land Command and Control Evolution) for the Italian Army [10].

In so doing, the paper is organized as follows: Sect. 2 underlines the main characteristics of the Command and Control function and why developing in terms of mission threads is convenient; Sect. 3 presents some details about the one-team system on which the IAA has been based; Sect. 4 discusses how IAA has grown and how the portfolio management issues have posed a threat to the entire process but also highlights how solutions have been implemented; and finally, Sect. 5 draws the conclusions.

2   Command and Control

Any Command and Control (C2) capability is essentially based upon the exercise of authority and direction by a properly designated commander over assigned forces in the accomplishment of the mission [9]. The challenge is to deliver this capability in the form of a C2 "automated system" which seamlessly integrates hardware, software, personnel, facilities, and procedures. Furthermore, such system is to orchestrate and enact automated and efficient processes of information collection, processing, aggregation, storage, display, and dissemination, which are necessary to effectively exercise command and control functions [11].

LC2EVO is based upon the development of an automated process for managing mission threads (developed as Functional Area Services—FAS) which consist of an operational and technical description of the "end-to-end" set of activities and systems that enable the execution of a mission.

These mission threads are a useful mechanism to set up a framework for developing users' needs for a new system. They describe the system from the user's prospective as well as the interaction between one or more actors, may these actors be end users, other systems, or hardware devices.

Each mission thread is further characterized by specific variables, to include geographical/climate, cultural/social, and even military functional areas or single organization specificities.

Pursuant to NATO ISAF doctrine, the Mission threads on which LC2EVO development has been based upon are as follows:

- Battle space management (BSM),
- Joint intelligence, surveillance, reconnaissance (JISR),
- Targeting joint fires (TJF),
- Military engineering—counter-improvised explosive devices (ME-CIEDs),
- Medical evacuation (MEDEVAC),
- Freedom of movement (FM),
- Force protection (FP), and
- Service management (SM).

3   ITA Army Agile—The First Team

The first team (so-called IDT1) has been developing the first mission thread "battle space management" without any concern about the necessity of managing coherently more teams developing more MT (portfolio of products).

Since requirements change during the period between initial specification and delivery of a "piece" of the product, IAA appears to be very attractive for military applications, whereas the appetite of the stakeholder needs to be both stimulated and even educated. The concept supports Humphrey's Requirements Uncertainty

Principle, which states that for a new software system, the requirement (the user story) will not be completely known until after the users have used it [9, 12].

IAA, similar to SCRUM, allows for Ziv's Uncertainty Principle in software engineering, which observes that uncertainty is inherent and inevitable in software development processes and products [13]. And it accounts for Wegner's mathematical proof (lemma) that it is not possible to completely specify an interactive system.

IAA is thought to support changing requirements through the implementation of an incremental development approach and close interaction with the customer and the user and finally with all the stakeholders. According to this approach, the customer becomes part of the IDT, allowing a better understanding of the needs and a faster development of the solution.

The IAA methodology is not a one-size-fits-all approach in which all the details of the process are predefined and the development team have to stick with it without any modification. On the contrary, IAA defines a high-level approach and a "state of mind" for the team promoting change management and flexibility in the work organization aimed at satisfying the customer needs (which is the final target of the entire effort). As all the other AMs, IAA defines values, principles, and practices focused on close collaboration, knowledge sharing, fast feedback, task automation, etc.

An IAA team includes the following actors: a Product Owner Team (POT), a SCRUM Master (SM), an Integrated Development Team (IDT), and a IAA Coach.

The development process starts from a vision of the Product Owner (PO). Such vision is a broad stroke and high-level definition of the problem to address that will be refined and narrowed during the development through the backlog grooming. The backlog grooming is an activity that takes place during the entire development process focusing on sharpening the problem definition, pruning redundant and/or obsolete requirements, and prioritizing requirements. Moreover, the PO defines the scenarios and the criteria used to test (and accept) the user stories.

To adopt SCRUM for the purpose of developing software systems in close cooperation between the Italian Army and external contractors, the IAA team has been organized as follows [1]:

– **Product Owner Team (military)**: It provides the vision of the final product to the team. Members representing all the governance levels (according to the below-mentioned organization) are assigned to the POT. The more this role can be played by the stakeholder, the better it is, since shortening the length between the user and the developer has proved to be key for success. Embedding the stakeholder within the team, making him/her well aware of the methodology and providing for the ability of interfacing directly with the developers, is a way of pursuing customer satisfaction to the greatest extent. The availability of the stakeholder in fully supporting the IDT and the SM has a significant impact on the development itself serving concurrently as a vehicle for requirements' refinement and as a ground for building a sense of ownership of the final product which, in turn, is deemed to facilitate operational and users' acceptance [12]. For these reasons, the IDT members need to possess a wide knowledge about the system under development and a clear vision on the expected results. Moreover, they have to be involved in the testing of the system in accordance with a specific organization that the Army has developed. The contribution of the PO is of extraordinary importance, particularly at the beginning of the project during the definition of the product backlog and during the reprioritization activities required in case of significant changes in the backlog. Within the POT, the individual delegated to support the everyday life of the DP is the Operational Product Owner (OPO). The OPO participates actively in the development and in promoting team-building activities in conjunction with the SM.

- **IAA Master or simply Scrum Master (SM) (military)**: He is an expert in the usage of the IAA methodology and has a central role in the team coaching the IDT and helping in the management of the product backlog. The SM helps the IDT to address the issues in the usage and adaptation of the IAA methodology to the specific context in which the team works. Moreover, he leads the continuous improvement that is expected from any agile team. An additional duty of the SM is to protect the IDT from any external interferences and lead the team in the removal of obstacles that may negatively impact productivity. Once again the SM needs to be a military and to have a stronger attitude than his homologue in the "classic" SCRUM methodology. He is still a facilitator, but he does need to take the control of the team, accounting for the different cultural origins of the members. The SM and the PO are to closely and continuously collaborate to clarify the requirements and get feedback. He also has to take responsibility for overall progress and initiate corrective actions when necessary. Moreover, he is expected to build and sustain effective communications with key staff and organizational elements involved in the development as required.

- **Integrated Development Team (both military and civilians from contractors)**: Team members are collectively responsible for the development of the entire product (step by step, piece after piece), and there are no specialists focusing on limited areas such as design, testing, and development. All the members of the team contribute to the entire production. The DT is self-organized, although coming from different companies and even some of them from military branches, they are to be empowered to determine without any external constraints how to organize and execute the sprint backlog (the selected part of the product backlog that has been identified to be executed in a sprint or development iteration/cycle) based on the priorities defined by the POT. They are guided by the SA, who, as said, is a military in charge of managing the team in terms of problem solver, decision supporter, and to interface with the external complex organization. The DT usually includes between 3 and 5 people, all expert software developers.

- **IAA Coach (civilian from the main contractor)**: He/she manages and plans required resources within the framework of the IAA in accordance with the financial regulations and in the framework of the running contracts (it is not the core of this paper looking into the contractual issues). Industry is delegated authority to tailor expert knowledge to meet specific circumstances and coordinating pool of resources. Scrum Masters are in charge of directly seeking resources from IAA Coach, who is entitled to allocate them and negotiate the required time as well as the relevant labor profiles.

The goal of IAA is to deliver as much quality software as possible within each four-weeks-lasting sprint. Each sprint is to produce a deployable system, although partial, to be fully tested by the user for the purpose of providing fast feedback and to fix the potential problems or requirements misunderstandings.

In the context of the Army's activities, testing vests a primarily role and will be done at three different layers: (1) performed by the development team in the "development environment" and then in a specific "testing environment" having the same characteristics of the "deployment environment"; (2) testing activities performed by military personnel involved in test bed-based activities, in limited environments to validate the actual effectiveness of the developed systems in training and real operating scenarios (integrated test bed environment); and (3) testing activities "on the field" performed to verify the compliance of the developed systems to the national and international standards and gather operational feedback to improve the system's performance and usability [1].

IAA is characterized by short, intensive, daily meeting during which team member inputs system two variables (what has been developed the day before and what is to be performed on the specific date) for each active task.

At the end of each sprint, a sprint review takes place to verify the progress of the project by comparing development performance to predeveloped benchmark expectations. The results of the review are used to adapt the product backlog modifying, removing, and adding requirements. This activity is very important for the success of a project and involves all the member of the IDT and the interested stakeholders. The product backlog is dynamic, changing continuously collecting all the requirements, issues, ideas, etc., and should not be considered as a static elements as indeed it is designed to support the variability of the environment.

The sprint review is followed by the sprint retrospective in which the POT, the SM, and the IDT analyze together the sprint end results to evaluate its effectiveness and to identify issues and opportunities.

In conclusion, the content of the paragraph states that IAA has demonstrated having the organization to deliver quality software on time, focusing on the value from the customer's perspective [1, 4, 5] but limited to one IDT.

4   ITA Army Agile—TP2S (Team–Product–Portfolio–Strategy)

Critical to the overall development is a unified organization for managing growing complexity, involving more teams developing user stories for the whole C2 community, within several mission threads (Fig. 1). Starting from the just-one-team

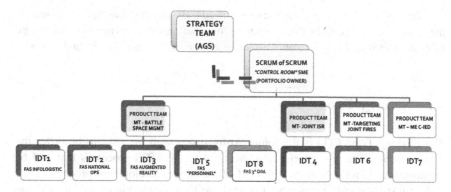

Fig. 1 TP2S governance

experience, noting the complexity growth resulting from a multi-team dimension, a Team–Product–Portfolio–Strategy (TPS2) governance has been set up based upon all the different levels of the military hierarchy. Each mission thread has been associated at least to one IDT. Sometimes (such as the team 1), the relation is one to many resulting in the need for product (equal mission thread) management. Coherence among all the mission threads is given by the portfolio management guided by the strategic level.

TP2S guides decision makers to strategically prioritize, plan, and control portfolios. It also ensures that the organization continues to increase productivity and iterative incremental product deliveries—adding value, strengthening performance, and improving results.

It eliminates surprises providing military hierarchy with a process to identify potential issues earlier in the development life cycle, to enhance the visibility hence to enable corrective action to be taken before final results are impacted.

It builds contingencies into the overall portfolio: Flexibility often exists within one IDT, but by integrating developments across the entire portfolio, organizations can achieve flexibility around how, where, and when it is needed to allocate resources, alongside the flexibility to adjust those resources in response to a crisis coming from an IDT.

TP2S does more with less: According to the figures resulting from the comparison between the classic development methodologies and IAA, customer satisfaction has been consistently achieved while cutting out inefficiencies, reducing costs, and ensuring a consistent approach to all products and portfolios.

It ensures informed decisions and governance by bringing together all team collaborators and processes into a single, integrated solution. A unified view of the portfolio status can be achieved within a framework of rigorous control and governance to ensure all products consistently adhere to business objectives.

TP2S extends best practice Army-wide: Organizations can continuously improve management processes and capture best practices, providing efficiency and customer satisfaction as a result.

Eventually, it forecasts future resource needs: By aligning the right resources to the right team at the right time, organizations can ensure individual resources are fully leveraged and requirements, in terms of user needs, are clearly understood.

4.1   Strategy Team

The Strategy Team (ST) is the senior Army Board with the responsibility for the development of the user stories and oversight of the TP2S processes that will be employed to exercise high-level control of the whole development process.

It will support POT (a member is part of the POT as well) in managing both the product backlog and the Product/Portfolio Teams to deliver iterations of the product to customers while also pursuing overall system coherence.

It will resolve issues arising from portfolio management and will identify IAA process high-level shortfalls and initiate efforts to address them.

4.2   Portfolio Owner—Scrum of Scrum (POSoS)

Individual teams work together to deliver working software at each iteration while coordinating with other teams within the program.

But, when several scrum teams work together on a large project, scrum of scrums is the natural next step for scaling agile, also in IAA framework. Firstly, the stand-up meeting component of scrum of scrums is a multi-team stand-up. This is *not* a status meeting nor is it a meeting for scrum masters to talk about agile process. It is a rather a short meeting to keep people from around the organization abreast of important issues across the portfolio and it touches on different dimensions in accordance with the technical expertise of the participants.

To get started, a member from each team has to be selected as a representative at the scrum of scrums in the foreseen format, ideally someone in a technical role. The scrum of scrums is a democratic meeting. A scrum master can help facilitate the stand-up, but unlike in single team's stand-ups, the aim here is to facilitate problem solving. As depicted in Fig. 2, the meeting will be held weekly, but alternatives are being arranged to enable the meeting to be held daily.

Nevertheless, whenever regardless of its cadence, this meeting is intended to open doors for knowledge sharing and to bring to the surface important integration issues as well as to serve as a forum for technical stakeholders to be made aware early and engage.

So far it has been chosen to have the scrum of scrums only once per week but run the meeting a bit longer than 15 min. The focus remains onto bringing issues affecting the group to the surface and determine what actions (if any) need to be taken and by whom [9, 14].

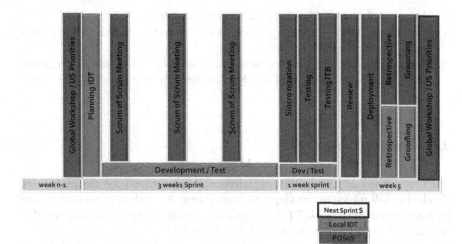

Fig. 2 SoS organization in IAA

IAA also finds value in extending other agile ceremonies such as sprint planning and sprint retrospectives to the scrum of scrums. Representatives meet just prior to their respective sprint planning and share what is likely to be pulled into their upcoming sprints (Global Workshop). This is a great way to avoid blocking dependencies between teams and address integration issues before they become cripplingly painful. For retrospectives, the scrum of scrums meets after their local team retrospectives and discuss action items that may require cross-team coordination to resolve.

While scaled-up planning and reflection may not need to be done every sprint, these *are* important parts of agile culture.

The POSoS manages the activity of the Product Teams (PTs) to deliver the overall project to time, cost, and user stories accomplishment.

When appropriate, it may refer issues to the ST for resolution where fundamental policy or strategy guidance is needed. He is also responsible for managing portfolio risks, dependencies, and customer expectations.

Within the overall context, the Portfolio Owner reports to the ST when and if needed (management by exceptions). Required tasks are discussed, agreed, and supported between the Portfolio Owner and the PT. Thereafter, day-to-day ad hoc direction shall be given directly by the Portfolio Owner, to ensure specific tasks progress to timely completion and to monitor costs and user stories accomplishment.

The POSoS is also entitled to supervise sprint planning, including the preparation of sprint backlog for each PT.

In exercising these responsibilities, POSoS, through the Global Operational Product Owner (GOPO), must ensure the achievement of coherence among constituent PTs including where appropriate:

- Management and coordination of user stories capture;
- Everyday activities, including support to architecture development and design with a reverse engineering process; and
- Management oversight of test and synchronization of system increments, including the establishment of suitable facilities where testing "on the military field."

4.3 Product Team (PT)

A PT is defined as a temporary organization that is created for the purpose of developing SW related to one single Mission Thread. More development teams can be incorporated into a single PT. If the team developing one single mission thread is unique, then the PT can be consolidated with the DT.

It ensures that everyone involved knows the expectations and contributes in keeping cost, time, and risk under control.

Successful teams have active management of risks, issues, and timely decision making.

The prime responsibility of the PT is to ensure that the project produces the product as depicted by the related user stories, to the required standard of quality and within the specified constraints of time and cost.

5   Conclusions

In this paper, an overview of the methodology for addressing the new Italian Army C2 system development process and the related governance process has been provided. The depicted analysis has started less than one year ago, but it has provided the Italian Army with a product that has met users' needs and expectations. Cost has also been significantly reduced. Although clearly a lot of work remains to be done, the inception of the initiative has produced significant results.

References

1. Cotugno FR, Messina A (2014) Adapting SCRUM to the Italian army: methods and (open) tools. OSS 2014, the 10th international conference on open source systems, Springer, Berlin
2. Sillitti A, Ceschi M, Russo B, Succi G (2005) Managing uncertainty in requirements: a survey in plan-based and agile companies, 11th IEEE international software metrics symposium (METRICS 2005), Como, Italy, 19–22 September
3. Schwaber K (2004) Agile project management with scrum, Microsoft Press, USA
4. Ruggiero M (2014) Coalition information sharing and C2 & enterprise systems evolution @ Italian army, briefing given at the NATO ACT strategic command, Norfolk, 10 December 2014

5. Origin A (2004) Method for Qualification and Selection of Open Source Software (QSOS), http://www.qsos.org
6. Wasserman A, Pal M, Chan C (2005) Business readiness rating project. BRR whitepaper. http://www.openbrr.org/wiki/images/d/da/BRR_whitepaper_2005RFC1.pdf
7. Jermakovics A, Sillitti A, Succi G (2013) Exploring collaboration networks in open-source projects. 9th international conference on open source systems (OSS 2013), Koper, Slovenia, 25–28 June
8. Messina A (2014) Adopting Agile methodology in mission critical software production. Point paper. http://ec.europa.eu/informationsociety/newsroom/cf/dae/document.cfm?
9. Sutherland J (2001) Agile can scale: inventing and reinventing SCRUM in five companies. CUTTER IT J 12(14)
10. Cotugno FR, Messina A (2014) Implementing SCRUM in the army general staff environment, SEDA 2014, the 3rd international conference in software engineering for defence applications
11. U.S. DOD (2008) Interoperability and Supportability of Information Technology and National Security Systems (CJCSI 6212.01E)
12. Sillitti A, Succi G (2005) Requirements engineering for agile methods. In: Aurum A, Wohlin C (eds) Engineering and managing software requirements, Springer
13. Ziv H, Richardson D (1997) The uncertainty principle in software engineering, 19th international conference on SW engineering, IEEE
14. https://www.atlassian.com/agile/ways-to-scale-agile, 07/04/2015 18:00
15. Humphrey WS (1996) Introduction to the personal software process, Addison Wesley, Boston
16. http://en.wikipedia.org/wiki/Project_portfolio_management

How Agile Development Can Transform Defense IT Acquisition

Su J. Chang, Angelo Messina and Peter Modigliani

Traditional defense acquisition frameworks are often too large, complex, and slow to acquire information technology (IT) capabilities effectively. Defense acquisition organizations for years have been concerned about the lengthy IT development timelines and given the pace of change in operations and technology, it is critical to look for new strategies to acquire IT for defense systems. Over the last decade, agile software development emerged as a leading model across industry with growing adoption and success. Agile is centered on small development Teams delivering small, frequent releases of capabilities, with active user involvement. From a planning and execution viewpoint, agile emphasizes an iterative approach with each iteration informing the next. The focus is less on extensive upfront planning for entire programs and more on responsiveness to internal and external changes, such as operations, technology, and budgets. Based on US and Italian experiences, this paper discusses some of the common challenges in implementing agile practices and recommended solutions to overcome these barriers.

Angelo Messina: Deputy Chief Italian Army General Staff Logistic Department and DSSEA Secretary.

S.J. Chang (✉) · A. Messina · P. Modigliani
The MITRE Corporation, Virginia, USA
e-mail: sjchang@mitre.org

A. Messina
e-mail: angelo.messina@esercito.difesa.it

P. Modigliani
e-mail: pmodigliani@mitre.org

© Springer International Publishing Switzerland 2016   13
P. Ciancarini et al. (eds.), *Proceedings of 4th International Conference
in Software Engineering for Defence Applications*, Advances in Intelligent
Systems and Computing 422, DOI 10.1007/978-3-319-27896-4_2

1 Introduction

Agile software development practices integrate planning, design, development, and testing into an iterative lifecycle to deliver software at frequent intervals. Structuring programs and processes around small, frequent agile releases enable responsiveness to changes in operations, technologies, and budgets. These frequent iterations effectively measure progress, reduce technical and programmatic risk, and respond to feedback and changes more quickly than traditional waterfall methods.

While the commercial sector has broadly adopted agile development to rapidly and dynamically deliver software capability, agile has just begun to take root across many defense organizations.[1,2]

Agile development can be distilled into four core elements:

- Focusing on small, frequent capability releases;
- Valuing working software over comprehensive documentation;
- Responding rapidly to changes in operations, technology, and budgets; and
- Active users involved throughout the development process to ensure high operational value.

The foundation of agile is a culture of small, dynamic, empowered Teams actively collaborating with stakeholders throughout product development. Agile development requires Team members to follow disciplined processes that require training, guidance, and openness to change.[3] While agile does impose some rigor, the method does not consist of simply following a set of prescribed processes, but instead allows dynamic, tailored, and rapidly evolving approaches that suit each organization's IT environment.

The first section of this paper will discuss the Italian Army (ITA) experience with agile implementation and how they have managed the hurdles related to process and culture. The second section of the paper will discuss agile in the US DoD acquisition environment, the prerequisites for agile adoption in the USA, and the barriers as it relates to program structure, requirements, and contracting. Lastly, the paper will summarize with a comparison of both the ITA and DoD experiences with agile adoption.

2 Agile in the Italian Army Environment

The lessons identified and learned for the last ten years of military operations in the area of the Command and Control software have clearly shown the limits of the software engineering traditional approach. The rapidly changing scenario of every

---

[1]Lapham et al. [10].
[2]Northern et al. [15].
[3]US Government Accountability Office [8].

operation has pointed out how difficult it is to define a stable and consolidated user requirement for the C4I software applications. The ITA General Staff call it as "volatility of the user requirement." On the other end, the software products to be used in the military applications are traditionally developed in strict adherence to software quality standards to limit the risk produced by errors, faults, and malfunctions.

At the ITA General Staff, it was decided to try to overcome the problem of the requirement volatility by substituting the physical document with a "Functional-technological Demonstrator" a software prototype to be used by the military user to describe his needs in practical terms by interacting with the proto-type. The "agile" methods seemed to be best candidates to this particular kind of production. The SCRUM declination of agile was the selected approach because of the short and fixed length of the "Sprint" production cycles and the clear definition of the roles.

When the ITA Team was at the point of starting this new approach for the first time, there was a wide spread awareness in the military community of the risks connected to the agile methods: Scrum, XP, Kanban, etc. Nonetheless, the need for a paramount change was such to be willing to take the risk. This is the first step every organization must face: taking the risk but managing it deliberately and aggressively.

At the time the first Team was started, agile scrum had been around as long as almost 10 years and though there were many success stories, there are as many that can also be characterized as a failure. As a matter of fact, when the first Team finished the first Sprint and there was a "real" delivery, it was quite a surprise. The people in charge of the effort where quite ready to watch a couple of "warm up pseudo-sprints" but to their surprise, all worked pretty well since the very beginning.

In the initial period, a technical investigation or a comparative analysis with other similar efforts was out of the scope of the initiative. Later on, the need to come to a thorough understanding of why this Scrum-derived method worked so well in the Army Logistic environment became a real need. After the first release of the software realized with this new approach in a seven-month period, the need to understand the reasons of such a good performance was no longer a methodology issue or a software engineering curiosity but a real necessary step in the process of consolidating a production procedure.

After a year in the process, the ITA has seven agile Teams in place and working. The full "Scrum of Scrum" complex production line is active and synchronized. The Teams turned on progressively taking charge of the full architecture of an overarching strategic Command and Control tool that implements full spectrum of the decision support activity, stretching from the facilities management to the operational Field Application Services (FAS).

The project was named LC2EVO and is now working on a 5-week "Sprints" delivery cycle.

3 Italian Army Culture and Processes

Effective agile adoption requires a new culture and set of processes that may be radically different from those in many defense organizations today. Agile practices, processes, and culture often run counter to those in the long-established defense organizations. The agile model represents a change in the way the government conducts business, and programs must rethink how they are staffed, organized, and managed, as well as whether the business processes, governance reviews, funding models, that support an acquisition are structured to support agile. The following eight reasons are why early adopters in ITA Army defense organizations are succeeding.

1. **Trust in people**. The empowerment of the Team should be deeply enforced from the beginning. In the case of the ITA, it was not easy nor simple, but the first step in shaping the agile effort was to value individual skills and commitments. A scrum tool was used for this purpose: the use of a scrum dashboard with "post-its" and all the relevant connected "ritual" processes was used by the ITA. The scrum "Champion" and his co-workers took full responsibility for the result of the effort but implemented a zero-tolerance acceptance of all the scrum elements the Team accepted to implement. In particular, some psychological side effects were actively pursued. In the case of the scrum dashboard for example, the electronic equivalent of the "post-it" user story was implemented in parallel to ensure traceability and documentation, but it never substituted the physical act of drawing the "post-it" with the programmer name and sticking it on the "to do" area of the dashboard. The value of this simple (public) ritual has deep implications to gain personal commitment.

2. **People do their best if given enough freedom**. An integrated development Team with subject-matter experts (SMEs) working together with both contractor's and government analysts and programmers is the optimal Team mix. The pressure generated by the direct relationship with the product owner (PO) is useful to inspire strong work commitments and create positive tension. The SMEs benefit from actually witnessing "their" functionality take shape and become working software; and the programmers take pride in their ability to translate user requirements into software functionality. When the right level of performance or quality is not being achieved, there is shared accountability for the success or failure of the results that force the Team to work together on common solutions.

3. **No project management on top of Scrum Teams**. According to the ITA agile methodology, putting program management on top of a "Scrum Team" is a conceptual mistake. Sprints are timeboxed and the only date available for public release is the end of the Sprint. Any question such as "are you able to finish by xx xx?" has to be rephrased as "How many stories are you able to finish within the time box of the next Sprint?" This is probably the toughest issue to solve with any legacy organization. Flowcharts, pert charts, and time-marked programs are incompatible with this vision of agile scrum.

4. **Scrum does not improve software quality, capable people do**. Having low-skilled personnel on the Team is not acceptable. For some professionals (security, data architecture, software architecture, and few others) top-level expertise is required. Sometimes, teamwork and a good Scrum Master can work out situations where there is a lack of specific expertise. In such cases, "technical gap filling training" stories have to be added to the product backlog and their priority level has to be acknowledged by the product owner/stake holder.

5. **An agile Team continuously improves**. That is why Scrum has retrospectives to see what went well, what can be improved, and to define actions. People have a natural tendency to seek improvement. In the ITA agile "scrum" environment, improvement is almost automatic: It comes out by the continuous confrontation process in the Team. Each Team member learns from the others especially from the non-programmers.

   Resistance to change is one of the major issues to overcome when implementing any agile production Team. Resistance to the new way is almost as natural as the tendency to seek continuous improvement. In the LC2EVO implementation, not much of this problem was found among the analysts and programmers, but it could have been an issue among the management. As stated before, no traditional program management of any kind was allowed in the ITA Army agile effort.

6. **The product owner role**. The PO is never alone. A product ownership board is recommended that it includes stake holder representatives with decision capability. A PO has to continuously work with the Team to ensure the "nominal" scrum functions are met and give the Team a precise point of reference. On the other end, in the ITA vision, a PO has to keep close contact with the stake holder to refine the user vision of the evolving "requirements" and make sure the stake holder's expectations are met. The adoption of a PO board ensures a continuous link with both sides of the community: the stake holders and the Team. Being part of the Team (at least with one member), the PO board is fully aware of the level of quality of the developed software. In the domain of "mission-critical" applications, the quality and then the security and reliability of the product are specified by the definition of high priority user stories (written by the relevant experts on the Team) and the implementation of such stories cannot be different. It is true that the stake holder side of the PO board will try to go back to the old way "Just deliver those features as fast as possible" and will try to transform the Sprint delivery date in a deadline but this can be avoided with the continuous watch by the POs and Scrum Master on the correctness of the approach.

7. **Product quality**. A controlled environment where the use of coding standards was the rule was implemented by the ITA Army. To avoid "cowboy coding," the ITA Army agile environment includes a layer of software tools which give in real time the quality level of the produced code. This environment is, first of all, a support to the programmers to understand how well they are performing. In the event the level of quality decreases, they have the possibility to insert "knowledge acquisition" stories in the product backlog to fill "technical or cultural" gaps.

8. **Tailored organization**. As stated before, no project management on the stakeholder side is accepted as the "fee" to pay in exchange of the possibility of modifying the product backlog after any of the Sprints. Any negotiation is carried on in terms of features (user stories to be deleted, inserted or changed in priority) to be delivered in the next Sprint(s).

The matter of the contractual implementation of the agile framework is far from being solved and many groups are working on this issue. In the ITA Army agile effort [3] to produce LC2EVO, a relevant decrease in the total cost of the produced software was experienced. Detailed investigation on the technical details of this decrease is still ongoing, but at first glance, it is reasonable to say that the main cost reduction drivers were the following:

- Only the required documents were elaborated, no one was tasked to write useless pieces of paper that nobody was going to read just because of a standard.
- The focus by the Team on the most desired features ends up in a "natural" code density in the areas most valued by the customer producing some kind of multiplying factor (still under investigation).
- User stories with no business value are quickly dropped out and canceled from the product backlog. In this area, the quick delivery feature of the agile methods helps to stimulate the customer focus on what he really wants which may not be so clear at the beginning of the project.

4 Agile in the DoD Environment

Despite the success that agile development has achieved in the private sector, commercial implementation of agile does not directly translate to agile adoption in the DoD environment. The barriers to program structure, requirements, contracting, and culture and processes often stem from these key differences. First, the government must adhere to a set of rigorous policies, statutes, and regulations that do not apply to the same degree to the commercial sector.[4] Following the rules that govern federal acquisition often involves a bureaucratic, laborious, and slow process that greatly influences how effectively DoD can implement agile. Second, the commercial sector has a different stakeholder management process than the government. Private firms are accountable to an internal and layered management structure that usually goes no higher than a corporate board of directors; the few possible external stakeholders (e.g., labor unions) rarely cause frequent and major disruptions.

The government bureaucracy has layers upon layers of stakeholders with a high degree of influence that can create frequent and significant disruptions. Everything from a change in the political administration to budget sequestration can exert

---

[4]Lapham et al. [11].

significant external influence on a DoD program. Lastly, the bureaucratic layers of government make it difficult to empower agile Teams to the same extent as in the private sector. The commercial sector has considerable latitude to make adjustments throughout the course of the development because companies closely link accountability, authority, and responsibilities to push decision making to the lowest levels. The government's tiered management chain of command makes it difficult for the agile Team to make decisions quickly and unilaterally.

The above comparisons demonstrate the need for DoD to tailor agile processes to its unique set of policies and laws. Herein lies the fundamental issue with agile adoption in DoD. The practices, processes, and culture that have made agile development successful in the commercial sector often run counter to the current practices, processes, and culture in the long-established defense acquisition enterprise.[5] In many ways, the acquisition environment needed to execute agile development is the opposite of the acquisition environment in place today.

- The small, frequent capability releases that characterize the agile development approach directly contrast with the traditional DoD acquisition model designed for a single big-bang waterfall approach.[6] Currently, every step in the acquisition system must be extensively documented and approved prior to execution. For example, according to DoD policies, an IT acquisition program must meet 34 statutory and regulatory documentation requirements prior to beginning development,[7] whereas agile emphasizes working software over comprehensive documentation.[8]
- Agile also enables rapid response to changes in operations, technology, and budgets. By contrast, DoD requires budgets, requirements, and acquisitions to be planned up front, often several years in advance of execution, and changing requirements, budgets, and strategies during the execution process is disruptive, time-consuming, and costly.[9]
- Lastly, agile values active involvement of users throughout the development process to ensure high operational value and continuously re-prioritizes the ongoing requirement process on the basis of feedback from the user community on deployed capabilities. Today's DoD requirement process is static, and rigid and limits active user involvement and feedback during the development process.[10]

---

[5]Broadus (January–Feburary [2]).
[6]Ibid.
[7]Defense Acquisition University (DAU) [4].
[8]Lapham, DoD agile Adoption, [9].
[9]Modigliani and Chang (March [14]).
[10]Lapham et al. [10].

Given the above key differences, DoD has been ill prepared to adopt agile development practices and in fact agile implementations so far have not always succeeded. Some early DoD adopters attempted what they thought or promoted as "agile," yet they did not incorporate some of the foundational agile elements into their structures or strategies. This resulted partly from the lack of definition and standardized processes for agile in the federal sector. In some cases, programs implemented a few agile principles, such as breaking large requirements into smaller increments, but did not integrate users during the development process to provide insight or feedback. Other programs structured capability releases in a timeboxed manner,[11] yet did not understand what to do when releases could not be completed in time.

Adopting only a handful of agile practices without a broader agile strategy often fails to achieve desired results.[12] For example, one DoD early adopter initially attempted to implement agile practices by breaking large requirements into several 4-week Sprint cycles. However, the program lacked high-level agreement on what to develop in each cycle and did not have a robust requirements identification and planning process in place. Furthermore, the program lacked an organized user community and active user-participation throughout the development process—a fundamental agile tenet. As a result, the agile processes quickly degenerated and the program only delivered 10 % of its objective capability after two years of failed agile development attempts. The program finally retreated to a waterfall-based process. It simply could not execute the agile strategy without the proper environment, foundation, and processes in place. On the other hand, DoD has recorded some significant successes with agile, such as the Global Combat Support System–Joint (GCSS-J) program, which has routinely developed, tested, and fielded new functionality and enhanced capabilities in six-month deployments.[13]

Another successful early agile adopter had their testing leaders report "after a decade of waterfall developed releases, the first agile iteration is the most reliable, with fewest defects, than any other release or iteration." They attributed it to the frequent feedback, number of releases, continuous builds, and engineering releases with stable functionality. DoD programs are beginning to turn the corner to successfully adopt agile practices. The programs that have had the most success were ones with major capabilities already deployed and used agile practices for subsequent upgrades. The goal is to position IT programs to adopt agile from the start of the acquisition life cycle.

---

[11]A timebox is a fixed time period allocated to each planned activity. For example, within agile, a Sprint is often timeboxed to a 4- to 6-week-time period or a release is timeboxed for a 4- to 6-month time frame.

[12]US Government Accountability Office [8].

[13]Defense Information Systems Agency [5].

5 Prerequisites for Agile Adoption

As a starting point, defense organizations should adopt a common understanding of agile and identify the underlying set of values that describe the purpose and meaning of their agile practices. The authors propose the following guiding principles for agile adoption:

1. **Focus on small, frequent capability releases to users**—Smaller releases are easier to plan, present lower risks, and are more responsive to changes. Projects should focus on delivering working software as the primary objective.
2. **Embrace change**—Projects must allow for changes to scope and requirements based on operational priorities, user feedback, early developments, budgets, technologies, etc. This requires flexible contracts, strong collaboration, and rigorous processes. Projects should plan early and then adapt based on current conditions.
3. **Establish a partnership between the requirements, acquisition, and contractor communities**—Projects should foster active collaboration on operations, technologies, costs, designs, and solutions. This requires committed users who contribute to development, tradeoff discussions, and regular demonstrations of capabilities. One DoD program had the acquirers, contractors, and users all participating in agile training. This provided a foundational education and a forum to discuss the specific agile processes for their program.
4. **Rely on small, empowered, high-performing Teams to achieve great results**—Organizing around each release with streamlined processes and decisions enables faster deliveries that are more successful. A DoD program benefited greatly with an integrated, high-performing Team by co-locating users, acquirers, testers, requirements leads, and many developers.
5. **Leverage a portfolio structure**—Individual programs and releases can deliver capabilities faster by using portfolio or enterprise strategies, processes, architectures, resources, and contracts.

These tenets align with the recommended set of principles in the Government Accountability Office (GAO) report on "Effective Practices and Federal Challenges in Applying agile Methods."

6 Agile Requirements Process

The agile requirements process values flexibility and the ability to reprioritize requirements as a continuous activity based on user inputs and lessons learned during the development process. In contrast to current acquisition practices, the agile methodology does not force programs to establish their full scope, requirements, and design at the start, but assumes that these will change over time.

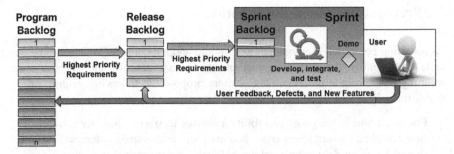

Fig. 1 Program, release, and Sprint backlogs

The program backlog contains all desired functionality and requirements. A release backlog typically comprises the highest priority requirements from a program backlog that a Team can complete within the established time frame. A Sprint then addresses the highest priority requirements from the release backlog. Once the development Team commits to the scope of work for a Sprint, that scope is locked. Sprint demonstrations conducted by the contractor at the end of a Sprint may identify new features or defects that the Team would add to the release or program backlogs. Figure 1 shows the relationships among the program, release, and Sprint backlogs.

One DoD program developed half-page work packages for the program backlog, each with a rough government estimate, some design context, and technical dependencies/interfaces. Users prioritized the work packages from an operational perspective and aligned epics or themes to mission threads. Each spiral had roughly 60–70 work packages.

The product owner, actively collaborating with users and stakeholders, is responsible for grooming the backlog to ensure the content and priorities remain current as Teams receive feedback and learn more from developments and external factors. DoD programs found having a single or multiple product owners was program dependent and found success with both models. Programs should consider the size and diversity of the user base and the operational stakeholders' empowered representatives. Users and development Teams may add requirements to the program or release backlog or shift requirements between them. The release and development Teams advise the PO on the development impacts of these decisions, while users advise the release Team about the operational priorities and impacts. To address a specific user story, the program must understand dependencies on existing or planned capabilities. Some programs may turn to a change control board to make some of the larger backlog-grooming decisions. The use of this requirements process can help set a DoD agile acquisition program on the right path for implementation.

As noted in the earlier example, the ITA stated that "volatility of the user requirement" was a considerable challenge in implementing agile. They substituted the physical document with a "functional-technological demonstrator." Much like the Italians, several DoD programs have also adopted similar approaches, using databases, Excel spreadsheets, or agile-based software tools to track changes to user requirements. The agile manifesto emphasizes "working software over comprehensive documentation." Plans should be developed by the Team, for the Team, to provide some level of consistency and rigor, but the level of documentation should not impede the Team's ability to focus on capability delivery.[14] The key is to do "just enough" documentation that describes high-level requirements to satisfy the institutionalized requirements processes, but using another type of requirements tool and process to track the detailed user-level requirements that are subject to frequent changes and reprioritizations.

7   Structuring and Tailoring for Agile

To effectively integrate agile principles into a rigid, bureaucratic defense environment one must proactively tailor program structures and acquisition processes. Agile experts working closely with executives and process owners across each functional area can collaboratively design an agile, new framework for acquiring IT capabilities. The foundational element is the structure of the program's releases and redesigning the key acquisition processes to support the release time frames. While many of DoD's large IT systems are structured in five year increments, they have historically taken over eight years to deliver IT capabilities. Programs should structure releases to deliver capabilities in less than 18 months with a goal of 6–12 months. The smaller scoped releases reduce risk and improve chances of successful deliveries. Small, frequent releases enable the program to be responsive to changes in operations, technologies, and budgets.

Many of DoD's early agile adopters have designed releases 6–18 months long further subdivided by Sprints. Those programs that had 18-month releases believe an 8- to 12-month release structure would have been better suited to their environment. During release planning prior to development, they finalize their requirements and designs for the scope of the release. Monthly Sprints provided working software that address the highest priority release requirements first and deliver to the government. Some programs required interactive demonstrations with users every two Sprints, while others required integration into a government testing and demonstration environment at the release midpoint. Many reserved the final Sprint to not have planned features, but instead finalize any incomplete features and rework based on testing and feedback of the previous Sprints.

---

[14]Modigliani and Chang (March [14]).

8   Contracting for Agile Development

Contracting for agile development has proven tremendously difficult for US and Italian governments. While commercial firms often rely on in-house staff to execute the agile practices, the government must obtain agile development through contracted support. Long contracting timelines and costly change requests have become major hurdles in executing agile developments and enabling small, frequent releases. Contracting strategies for agile programs must be designed to support the short development and delivery timelines that IT requires.[15]

Complex laws and regulations often drive long contracting timelines, defined requirements upfront, rigid government contractor interaction, and contractor selection based on the strength of technical proposal. These runs counter to agile's tenets of short delivery cycles, dynamic requirements, close Team coordination, and using an expert Team with demonstrated success.

To overcome many of these challenges, programs should use a service contract instead of a product contract. A service contract provides the program with greater flexibility to modify requirements along the development process, because it describes the people and time required to execute the development process rather than locking-down the technical details of the end-product deliverable. However, this strategy assumes the government is the lead systems integrator and is responsible for overall product rollout and delivery. If the government expects the contractor to act as the systems integrator, determine the release schedule, and be accountable for overall product delivery, then a product-based contract in which the government describes overall delivery outcomes and objectives is more practical. However, this scenario would make it difficult for the government to execute a true agile process, because changes to requirements in the course of development, or a change to the delivery schedule, will require a contract negotiation that could affect the agile process.

Lastly, the program should focus on the competition strategy to be used for the initial award as well as for follow-on task orders and awards. This will help determine how to scope the contract or task order for each contract action. In some cases, the program would benefit from bundling a set of releases into a single contract action to minimize the number of contract activities during the development process. However, the program should balance this against the need to maintain continuous competition throughout the program life cycle to keep rates low and receive the best value for products and services.

One DoD program used many contractors, each having their own agile processes, epics, stories, story point system, etc. This required the DoD to learn and operate via multiple approaches and proved to be a challenge for requirements traceability. This is a trade-off between competition and developer autonomy versus government integration and management complexity.

---

[15]See Footnote 14.

9 Summary

ITA and DoD have experienced many similar challenges with agile adoption; however, both governments have also witnessed the potential for great success using this development methodology. Both encountered similar challenges with introducing agile methods [16] because they are so different from long-held traditional development methods. The culture and processes are the biggest obstacles to overcome, especially as it relates to a requirements community that is characterized as dynamic and volatile. However, despite these challenges, both countries were able to break from tradition and in many cases, fully embrace the agile methodology to deliver working software in short capability drops.

The focus on iterative development and frequent capability deployments makes agile an attractive option for many defense acquisition programs, especially time-sensitive and mission-critical systems. However, agile differs so profoundly from traditional development practices that defense organizations must overcome significant challenges to foster greater agile adoption. Leaders cannot expect individual programs to tailor current acquisition processes on their own, because the complexities of the laws and policies do not lend themselves to obvious solutions, let alone accommodate processes so fundamentally different from current defense practices.

This paper has offered potential solutions to these key challenges in order to aid programs in laying a foundation for successful agile implementation. As agile adoption continues to take root and expand across programs, defense organizations would benefit from additional guidance and training to ensure consistent and pervasive success in agile IT acquisition.

References

1. Balter BJ (Fall 2011). Towards a more agile government: the case for rebooting federal IT procurement. Publ Contract Law J
2. Broadus W (January–Feburary 2013) The challenges of being agile in DoD. Defense AT&L Mag, 5–9
3. Cotugno F, Messina A (2014) Implementing SCRUM in the army general staff environment. The 3rd international conference in software engineering for defence applications—SEDA, Roma, Italy, 22–23 September 2014
4. Defense Acquisition University (DAU) (2015, March 20) Milestone document identification (MDID). Retrieved from DAU: https://dap.dau.mil/mdid/Pages/Default.aspx?ms=2&acat=2&acatsub=1&source=All&type=chart0
5. Defense Information Systems Agency (2015, March 20) GCSS-J. Retrieved from Defense Information Systems Agency: http://www.disa.mil/Mission-Support/Command-and-Control/GCSS-J/About
6. Defense Science Board (2009) Report of the defense science board task force on: department of defense policies and procedures for the acquisition of information technology. Retrieved from Office of the Under Secretary of Defense for Acquisition and Technology: www.acq.osd.mil/dsb/reports/ADA498375.pdf

7. Duquette J, Bloom M (2008) Transitioning agile/rapid acquisition initiatives to the warfighter. The MITRE Corporation, Virginia
8. U.S. Government Accountability Office (2012) Software development: effective practices and federal challenges in applying agile methods (GAO-12-681), U.S. Government Accountability Office, Washington, DC
9. Lapham MA (2012, January/February) DoD agile adoption. Retrieved from Cross Talk: http://www.crosstalkonline.org/storage/issue-archives/2012/201201/201201-Lapham.pdf
10. Lapham MA., Williams R, Hammons C, Burton D, Schenker A (2010) Considerations for using agile in DoD Acquisition. Software Engineering Institute, USA
11. Lapham M, Miller S, Adams L, Brown N, Hackemarck B, Hammons C, Schenker A et al. (2011) Agile methods: selected DoD management and acquisition concerns. Software Engineering Institute, USA
12. Messina A (2014) Adopting Agile methodology in mission critical software production. Consultation on Cloud Computing and Software closing workshop, Bruxelles, 4 November 2014
13. Messina A, Cotugno F (2014) Adapting SCRUM to the Italian army: methods and (open) tools. The 10th international conference on open source systems, San Jose, Costa Rica, 6–9 May 2014. ISBN 978-3-642-55127-7
14. Modigliani P, Chang S (March 2014) Defense agile acquisition guide. The MITRE Corporation, Virginia
15. Northern C, Mayfield K, Benito R, Casagni M (2010) Handbook for implementing agile in department of defense information technology acquisition. The MITRE Corporation, McLean, VA
16. Office of the Secretary of Defense (2010, November) A new approach for delivering information technology capabilities to the department of defense. Retrieved from Office of the Deputy Chief Management Officer: http://dcmo.defense.gov/documents/OSD%2013744-10%20-%20804%20Report%20to%20Congress%20.pdf

AMINSEP-Agile Methodology Implementation for a New Software Engineering Paradigm Definition. A Research Project Proposal

Vincenzo Mauro and Angelo Messina

Abstract The application of the "Agile" methodology to a Military Application Software production line is not to be taken for granted. The adoption of the new production style has implied the solution of multiple problems and the generation of a custom version of the basic "Scrum" theory. One of the major changes has been imposed by the need to include Military Test Beds and Military Units in the workflow. Even if the "Agile Scrum" methodology has been around for over a decade now, there are many successful implementation stories the Italian Army experience and clearly show there is more conceptual work to be done. There are a number of open issues to be tackled and a growth potential to be exploited. AMINSEP tackles the area of monitoring and measuring using noninvasive tools (Italian Army choice) necessary to keep high the product quality level and monitor criticalities. Those tools need to evolve. The evolution has to take into account the peculiar environment of the agile teams. The current tools are based on complexity metrics borne for the old fashion software factories. The collected data do not give a complete picture of the team-based code development cycle. A new set of metrics is needed to complement the existing ones. The basic concept and the structure of the research project are presented.

Vincenzo Mauro: Italian Secretariat General of Defence.

Angelo Messina: Italian Army General Staff.

V. Mauro (✉)
Italian Secretariat General of Defence, Rome, Italy
e-mail: vincenzo.mauro1@esercito.difesa.it

A. Messina
Logistic Department of Italian Army, Rome, Italy
e-mail: angelo.messina@esercito.difesa.it

© Springer International Publishing Switzerland 2016
P. Ciancarini et al. (eds.), *Proceedings of 4th International Conference in Software Engineering for Defence Applications*, Advances in Intelligent Systems and Computing 422, DOI 10.1007/978-3-319-27896-4_3

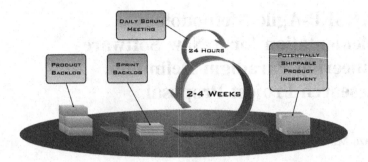

Fig. 1 Traditional "Scrum" approach

1 Introduction

The lessons identified and learned from the last ten years of military operations in the area of the Command and Control software have clearly shown the limits of the software engineering traditional approach. The rapidly changing scenario of every operation has pointed out how difficult it is to define a stable and consolidated user requirement for the C4I software applications. The Italian Army General Staff call it: "Volatility of the user requirement." On the other end, the software products to be used in the military applications are traditionally developed in strict adherence to software quality standards (i.e., mil 2167A, DOD 2167, ISO IEC 9126) to limit the risk produced by errors, faults, and malfunctions.

At the Italian Army General Staff, it was decided to try to overcome the problem of the requirement volatility by substituting the physical document with a "Functional-technological Demonstrator," a software prototype to be used by the military user to describe his needs in practical terms by interacting with the prototype. The "agile" methods seemed to be best candidates to this particular kind of production. The SCRUM declination of agile was the choice because of the short and fixed length of the "Sprint" production cycles and the clear definition of the roles. After one year, the "ITA ARMY AGILE" method has been defined as particular declination of agile, leveraging rules and rituals from Agile Scrum and integrating tools and procedures peculiar of the Army environment. This method seems to be particularly useful to implement production lines (teams) in the area of high-reliability mission-critical software (Fig. 1).

2 The Research Area

The application of the "Agile Scrum" methodology in the Italian Army software development experience has clearly shown a number of open issues to be analyzed. In order to face the related problems and exploit opportunities, deep investigations are needed in several research sectors.

2.1 Collection and Management of "User Stories"

Natural Language is an easy tool for users but besides its embedded ambiguity is too unstructured for developers to correctly figure out the correlated tasks. The need for some discipline in the process of collecting the user stories must not create prejudice for the human interaction that is the base of the effectiveness of the user stories collection (non linear process).

According to the "theory," the user stories must respond to the INVEST criterion (Independent Negotiable Valuable Estimable Short Testable). Most of the above attributes belong to the linguistic-semantic arena not to the SW engineering one. Some research on this subject is needed to include a human sciences expertise point of view on this part of the sprint planning. The use of some computational linguistic-based tool could be considered in parallel.

Once the user stories are collected, some kind of (formal) CASE tool has to be used as entry point to a SW engineering process. Using, for example, UML to "formalize" a Product Back Log is an unwanted form of translation. In such a case, there is a risk of loss meaning and effectiveness of the user stories. This is a topic for investigation.

2.2 Monitoring and Measuring

Noninvasive tools, preferred by the Italian Army, are necessary to keep high the product quality level and monitor criticalities. But continuous improvement and adaptation of those tools is necessary, taking into account the peculiar environment of the agile teams. Current tools are based on complexity metrics borne for the old fashion software factories: the collected data do not give a complete picture of the code development cycle implemented by the agile team: new metrics are, therefore, needed to complement the existing ones. Open question is: what is the right amount of "control" for the agile environment? The needs are to monitor and assess without disrupting the intrinsic nonlinear elements of the agile methodology, which are matching the correspondent ones embedded in the software product nature, giving much better results when compared to the linear, traditional, "waterfall" methods.

2.3 "Scrum" Adaptation

Although the Scrum methodology [1] is a framework flexible enough to adapt to different projects and processes (and adaptation itself is one of three pillars of the scrum theory), the peculiarity of the military environment requires some reflections and modifications on aspects usually considered strictly unchangeable.

One of the most critical aspects on this perspective is the application of the rule concerning the unicity of the Product Owner, defined as "the sole person responsible for managing the Product Backlog."

Fig. 2 Traditional "Scrum" team composition

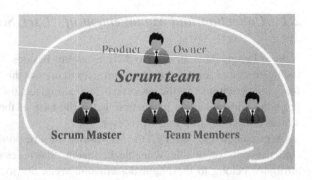

Product Owner is always accountable for expressing Product Backlog items, ordering the items in the Product Backlog to best achieve goals and missions, optimizing the value of the work of the development team, ensuring that the Product Backlog is visible, transparent, and clear to all (Fig. 2).

We will see later that in our experience the Product Owner is represented by multiple entities. This specification has a relevant impact on team composition and may introduce a degree of complexity also on scrum artifacts.

2.4 New SW Engineering Paradigm

Mission-critical SW has to be developed in a software engineering environment that ensure the quality of the product and enables all traditional features needed for the product lifecycle management. Unfortunately, all the available tools and the expertise themselves are legacy of the traditional methods and are not so effective in a scrum environment. Just to give an example out of the Army Scrum experience: software high-level design (architecture) in "agile," being very simplistic, is a parallel concurrent effort with the coding activity. Software architects, therefore, need flexibility and vision to merge their effort with the analysts/programmers one. Professionals familiar with the "traditional" approach are reluctant to accommodate changes in the architectural design during the scrum production cycles and tend to produce an exaggerated quantity of "nonfunctional requirements" to block the product backlog. Professional skills in the area (SW architects) with necessary flexibility are difficult to find.

3 Horizon 2020 Opportunities

Research activities in this area fully qualify for possible funding under the "Horizon 2020" European Union program (Fig. 3).

Fig. 3 Horizon 2020
framework programme

Information and communication technologies (ICTs) underpin innovation and competitiveness across private and public sectors and enable scientific progress in all disciplines. Thus in Horizon 2020, ICT-related topics can be found in all priorities, from "Excellence Science" to "Industrial Leadership" to "Societal Challenges." EU investments in ICTs are expected to increase by about 25 % under Horizon 2020 compared to the previous framework program, FP7.

Developments in ICT provide major opportunities for Europe to develop the next generation of open platforms on top of which a multiplicity of innovative devices, systems and applications can be implemented.

The "Software & Services, Cloud" unit recently undertook an open web-based public consultation on the research and innovation challenges for the Horizon 2020 Work Programme 2016–17 [2]. The challenges and research priorities identified in software (including Open Source) covered all phases of the software development cycle, starting with new approaches, languages and tools for requirements engineering; agile/scrum development frameworks; software development methodologies and tools for highly distributed systems; higher-level, intention-driven programming models; and approaches for the automation of interoperability and compliance testing, risk identification, detection, and mitigation.

Even if the "Horizon 2020" program is not intended to support pure military research projects, the "dual use" connotation of certain topics, such as software development, opens realistic opportunities for military sector.

This research project, in particular, is valid not only in the defense and security sector, but also in other areas where issues related to Command and Control

Table 1 Time schedule of the AMINSEP project (months dedicated to specific activities are gray shaded)

Month / WP	1	2	3	4	5	6	7	8	9	10	11	12	13	14	15	16	17	18	19	20	21	22	23	24
WP 0																								
WP 1																									
WP 2																									
WP 3																									

WP 0 - Coordination and Management of the Project
WP 1 - Assessment of the tools. Definition of the new model SW line and production of a model of the defectiveness of the SW
WP 2 - Implementation of the new model of the production line, definition and consolidation of optimized AGILE procedures
WP 3 - Case Study of a real Command and Control SW

software requirements and development have to be addresses in the framework of a complex organization and/or in critical missions (Table 1).

4   Aminsep

A research project proposal called AMINSEP (Agile Methodology Implementation for a New Software Engineering Paradigm definition) has been presented to the Secretariat General of Defence and National Armament Directorate of the Italian Ministry of Defence, with the purpose of obtaining a funding in the framework of the "National Military Research Plan."

The main proponent of the project is the Italian Army General Staff, that intend to develop the program in cooperation with the "Center of Information Systems of the Army," Academic Departments and Research Institutions.

The application of the methodology "Agile" to a "software factory for military" is not immediate and involved a series of adaptations of both procedures is of some roles that after a few months set up a variant "EI Custom" of the SCRUM methodology. The need for adaptation resulted mainly from the complexity of the structure that defines the end user, and therefore the Product Owner that in the case of systems C2 is represented by multiple entities: Army General Staff, Command for Education and Doctrine, and Military Units. The strength of the Scrum process E.I. is the opportunity to test the products (in an almost continuous cycle) on test bed already available (e.g., that of the "Center of Information Systems of the Army") and by Military Units, taking advantage of already planned exercises. These characteristics of Scrum E.I. configure specific methodology and determine the need to generate a conceptual framework for autonomous reference, enabling the formal definition of the model and the evolution of the production cycle. It is therefore necessary to compare the ideas and techniques born within Scrum E.I. with the current practice of software engineering. The approach was presented at the "Open Source Software Conference (OSS)" in Costa Rica on May 7, 2014 [3] and gained consensus on the reasons behind the choices made by the Italian Army, providing a positive feedback the validity of the methodology adopted, with the necessary variations, only able to handle the "volatile" requirement, also aimed at reducing costs, as well as user satisfaction and time compression.

It is now proposed to start a research on technologies which enables to determine SW tools, metrics, and everything else needed to consolidate a new concept (paradigm) of software engineering that allows a further reduction of the cost of the life cycle of the software and full coverage of the military needs of the user.

The project will be carried out in three phases:

I phase. Evaluation of tools and metrics used in a previous experience of the development team (software LC2-Evo Infra), fine-tuning of existing tools and inclusion of additional tools for the improvement of parameters of production (velocity, quality, defects, cost), with a goal of a minimum 20 % improvement.

Definition of the new model of SW production line. Definition of a model of the defectiveness of the SW.

II phase. Implementation of the new model of production line creating and/or integrating the development environment with COTS possibly Open Source; definition/consolidation agile procedures optimized for the new development environment; definition of a procedure for team building.

III phase. Implementation of a real case study through the creation of a computer procedure relating to a functional area of software support to the Command and Control indicated by Army General Staff; Final preparation of the analysis and the final demonstration.

5 Conclusions

Agile/scrum development methodologies can reduce development costs, increase customer satisfaction and adapt to volatile requirements changes. However, developing mission-critical and safety-critical systems with agile/scrum techniques requires further study, as the existing tools and expertise for the development of mission-critical and safety-critical software are not suited for agile development methods. Mission-critical SW has to be developed in a software engineering environment that guaranties the quality of the product and enables all traditional features needed for the product lifecycle management.

All the available tools and the expertise themselves are legacy of the traditional methods and are not so effective in a scrum environment [4].

Further work is therefore required on software monitoring and measuring systems and metrics as current tools are based on complexity metrics developed for traditional software development processes and are not well suited to modern agile/scrum frameworks.

The results of the AMINSEP research project will be applied and tested on a real case study, by developing a Command and Control support software chosen by the Army General Staff.

References

1. Schwaber K, Sutherland J (2013) The Scrum Guide. http://www.scrum.org. July 2013
2. Messina A (2014) Adopting Agile methodology in mission critical software production. In: Consultation on cloud computing and software closing workshop, Bruxelles, 4 November 2014
3. Messina A, Cotugno F (2014) Adapting SCRUM to the Italian Army: Methods and (Open) Tools. In: The 10th international conference on open source systems, San Jose, Costa Rica, 6–9 May 2014
4. Cotugno F, Messina A (2014) Implementing SCRUM in the Army General Staff Environment. In: The 3rd international conference in software engineering for defence applications—SEDA, Roma, Italy, 22–23 Sept 2014

Ita Army Agile Software Implementation of the LC2EVO Army Infrastructure Strategic Management Tool

Davide Dettori, Stefano Salomoni, Vittorio Sanzari, Dino Trenta and Cristina Ventrelli

Abstract The Army General Staff has identified the need for a strategic management tool for the numerous and widespread infrastructures belonging to the Italian Army. The generic requirement has called for various functions at different levels of information aggregation and fusion. The first attempt at finding a solution was based on the traditional software engineering approach which pointed out the difficulty of producing a high-definition requirement document. A customized "SCRUM Agile" software development methodology was then implemented producing excellent results. The process is described with particular focus on the User Community–Development Team relationship and the evolution of the associated Product Backlog. Results on the software cost reduction and on user satisfaction are reported. A detailed description of the User Stories evolution through the various Sprints is presented as a case study.

1 Introduction

Today, the army military infrastructures represent a real estate synthesis of the history of our Nation [1]. This is mainly due to an inheritance received at the end of the "Risorgimento" period or to substantial interventions which took place with the

D. Dettori (✉) · S. Salomoni · V. Sanzari · D. Trenta · C. Ventrelli
Stato Maggiore Esercito Italiano—IV Reparto Logistico, Rome, Italy
e-mail: davide.dettori@esercito.difesa.it

S. Salomoni
e-mail: stefano.salomoni@esercito.difesa.it

V. Sanzari
e-mail: vittorio.sanzari@esercito.difesa.it

D. Trenta
e-mail: dino.trenta1@esercito.difesa.it

C. Ventrelli
e-mail: cristina.ventrelli@esercito.difesa.it

© Springer International Publishing Switzerland 2016
P. Ciancarini et al. (eds.), *Proceedings of 4th International Conference in Software Engineering for Defence Applications*, Advances in Intelligent Systems and Computing 422, DOI 10.1007/978-3-319-27896-4_4

35

imminence of the conflicts of the First and Second World Wars. It is often a question of real estate property which was assigned to other uses and subsequently adequated to satisfy exacting allocations (Fig. 1).

From a recent reconnaissance of the Army's real estate, the patrimony is of about 3700 infrastructures with a total area estimated in 30,000 Ha.

Mainly because of political–financial reasons, the Army has been unable to face in a global or organic whole the problem of military infrastructures. For this reason, during recent years studies and sector projects have been elaborated which have evidenced the impossibility or, at least, the difficulties which have arisen in the endeavor to activate a specific national organization Plan.

Furthermore, the recent regulations in the alienation of the real estate of the Defense [2] have induced the Army to proceed with the rationalization of the available spaces of the non-strategic real estate, thus reducing management costs.

For this reason, a new Army infrastructure management Plan whose criterion guide centers on the new Defense Model [3] must be founded on an innovative Command and Control System [4, 5] which is able to face with unity and completeness the overall available information.

To finalize this Plan, an investigation has been carried out on the market for a new Informatic System which will be able to overcome the limits of current Defense data-base [6] to guarantee to the Army a innovatory software defined enterprise with direct access and updating capability via Web access by all the headquarters, departments, and organizations responsible for or concerned with Army real estate.

Moreover, not having and adequate know-how of the specific needs, a methodological and managerial approach of the military real estate was carried out through the identification of a high-quality functional software, suitable for the Army needs.

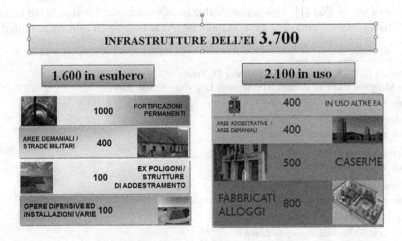

Fig. 1 Army real estate patrimony

The features searched were of a favorable product, not fully predictive, but rather adaptive in order to find suitable solutions to the problems of everyday life.

In this context, it is appropriate, from the beginning, to approach also the management of the real estate of the Army with a management development methodology able to provide an Agile, dynamic, and flexible product, using "System Integration & Infrastructure solutions" which have the following requirements:

- Portal (consisting of Core Services and F.A.S.[1]);
- Record;
- Identity and Access Manager;
- Infrastructure and Database;
- Virtualization and Cloud Computing;
- Storage and Backup Consolidation; and
- Security and compliance.

Consequently, a more modern and interactive development approach, defined "Agile," has been adopted within the Army (ITA ARMY AGILE [7]) and has been named "LC2EVO" [8].

"LC2EVO" has thus been identified as the new support system of Land Command and Control (LC2) based on fundamental services (core services) and on the different Functional Application Services (F.A.S.) focused on the direct administration of the various information sectors.

The first F.A.S. which was developed was a pilot project: It was the infrastructure F.A.S., named "Infra-FAS."

The "Agile" methodology [4] applied to the F.A.S. has foreseen the establishment of a miscellaneous team (*Scrum Team*) where specialists in software system design and development (*Development Team*) were supported by a Working Group (*User Community*) with an expert infrastructure management Officer and three architect and engineer Officers specialized in the real estate/building sector.

In this way, an osmosis between computer science and engineering may maximize the final output, focusing on the improvement of Command and Control management efficiency of the infrastructures of the Army.

In this specific case, through frequent exchanges of information and opinions within the above-mentioned development team, it was decided to start structuring a Infra-FAS, lasted for about 8 months, allowing in this way a continual check of the required software architecture and adapt them to the various needs which originated during the development and the gradual use.

In this article, a case study will be described through the different evolution phases and the prospective view of the user community (as one the involved stakeholders).

---

[1]Functional Area Service.

Starting from the real needs, it has been a success model where the unavoidable distorsious which exist between two universes, the digital and the real world, has been reduced to a minimum. In this way, the project idea was, as proved, as near as possible to the expectatious of the customers and the user (remaining community stakeholders).

2 The Development of the Project Seen by the "User Community"

At the start of "LC2EVO-Infra" project implementation, the complex typological, qualitative, and dimensional characteristics were analyzed, secondly, synthesizing the specific need in accordance with the following factors:

- **profound knowledge of the application domain** is to limit of the extrapolation mistakes following validation at the moment of release;
- **overall vision is** to predetermine the possible interventions for the perfecting of the project—for probing levels of the management complexity of the military infrastructures (reality)—also through the knowledge of the software already available in the Defense ambit;
- **analysis of possible variables** is to aim for a product which adopts itself to the changing needs of the final customer.

In substance, the elaboration process of the specific FAS-INFRA User Stories was carried out in 5 phases: definition, knowledge, comparison, projecting, and perfecting. These phases were necessary in order to involve a wide community of the consumers who were interested, in different ways, for the different competences related to infrastructures (Fig. 2).

In particular:

(A) **Definition**
   In this phase, the guidelines were defined shared by the Development Team for the whole subsequent infrastructure management process, prefigurating the ideal results ("what will this project do?," "toward who?," "with what aim?"). The elaboration of the common strategic vision was defined through a broad planning of the activities to perform (sprint planning meetings) to answer the continual requests of the Team stakeholders "component," with the aim to illustrate the following:

   - the general objectives to reach in the Sprint and the length of time needed to achieve them;
   - the hightest priority requirements contained in Product Backlog (Fig. 3).

Fig. 2 Preliminary activities for the preparation of the specific team Agile

Fig. 3 "Sprint Planning Meeting" example

(B) **Knowledge**

In this phase, the "Development Team" research and preliminary project activities, cooperating with the consumers who, above all, were more competent for each specific case (the cited infrastructural Working Group), to obtain a comprehensive vision, even if not complete, of the needs of the final users (at central level the Army General Staff for Command and Control activities and at a peripheral level, the Unit/Organization consignee of the specific property for the infrastructure).

The information obtained in this first phase was confronted:

- with the reality represented by the extrapolations (consequences) derived from the pre-existent Army infrastructure management systems, following the traditional methodology;
- with the application context to verify the suitability of the strategic vision.

During the knowledge phase, only the activities which were sufficient to give a first-quality delivery were studied and limited to:

- the interview of the members of the Team specialized in infrastructures;
- the collection of the most elementary identification requirements of the chosen properties (location, perimeters, registry, ...);
- the capability to confront and converse with the existing management software following the "classic method."

This choice is justified because of the capability to compensate possible gaps, during the completion phase, relative to the user and the context (Fig. 4).

(C) **Comparison**

Starting from the elements obtained during the first two phases, the Development Team has elaborated a first vision of the product which only the special consumer administrators (user community at central level) have been able to experiment.

The critical comparison with the stakeholders on these solutions has allowed a wide sharing of the project idea and of the product development in a medium term.

At the end of this phase, the characteristics of the first version of the product (models, behavior, conceptual interfaces) and the strategies for its release have been planned (Fig. 5).

(D) **Design**

During the design phase of the infrastructure management FAS a preliminary study was faced:

(a) of the graphical interface visual representation of every single significant Army infrastructure on a cartographical base, and classified as strategic, transferable, inactive) (Fig. 6);

Gestione Dati Tecnici Capacitivi

Nome

Artifact Content

User Story

Come utente di SME IV RL voglio poter tenere aggiornati i dati tecnici capacitivi relativi alle infrastrutture di F.A per censire la componenete logistica del patrimonio immobiliare dell'Amministrazione.

Business Value

Il censimento della componente logistica del patrimonio immobiliare di F.A. consente:

- il corretto monitoraggio dei costi di gestione
- il corretto monitoraggio di licenze/certificazioni/ idoneità delle varie capacità addestrative (Poligoni, CAGSM, etc) logistiche (Officina, Mensa, Magazzini etc)
- la disponibilità di alloggi ASC
- la tensione abitativa interna

User Story Elaboration

Le informazioni saranno aggregate in un documento propriamente detto Scheda Tecnica Capacitiva nel quale saranno raggruppati i seguenti parametri:

- capacità alloggiativa interna
- capacità direzionali
- capacità logistiche
- capacità sportive e ricreative
- capacità formative
- capacità addestrative
- costi di gestione
- servizi pubblici/sociali asserviti alla struttura

Fig. 4 User story 451—capacitive technical data management—20150507_0752734

(b) of the contents a selection of the essential descriptive and capacitive data (Fig. 7);

(c) of the dialog with the already-operating Defense General Staff Infrastructural Database (Fig. 8);

(d) of the insert mode of the selected infrastructures in the System and the subsequent population of data;

(e) of System communication (Fig. 9).

A "simplified" and "elementary" product of elevated potential of quantitative and qualitative evolution was obtained.

A minimum working high-technical-quality product was achieved which will serve as a base for subsequent developments during the refinement phase.

Fig. 5 Intermediate "LC2EVO" release

The project developed in this phase is the raw material in which the potential will be explored during the evolution and refinement phase.

The elaboration of the specific "user stories" in this phase is particularly complex because of the numerical consistence of the stakeholders involved, a number which is equal to 600 agencies, units and organizations, and for the diversified and inhomogeneous Army real estate patrimony—both in terms of architectural type (barracks, buildings, clubs, quarters, deposits, shooting ranges, logistical-training areas, communication sites, …) and in conservation state terms.

This was followed by a progressive insertion and validation of the most significant Army infrastructures in order to allow potential users to view the information by connecting to the "Infologistic"/"Infrastructure Management" section of the LC2EVO System, even if other functions were still in the implementation stages.

(E) **Perfectioning**
At the end of the preliminary phase of the project, and the following the intermediate release of the Product, the mixed Team worked parallely in the direction:

Fig. 6 LC2EVO-Infra F.A.S

- of a continual FAS implementation carried out by the Development Team
  component, reviewing the structural organization of the system;
- of a constant monitoring of the Product in a real utility context by the user
  community component;
- of the divulgation and indoctrination, through video conferences,
  explanatory briefings, help desk, messages and other, finalized for
  example:

  • to the communication of instructions concerning specific duties
    (Consignees, Technical Organizations of the Engineer Corps respon-
    sible for the territory, Army General Staff);

SCHEDA TECNICA CAPACITIVA						
DENOMINAZIONE INFRASTRUTTURA: _____						
E/D/R/C UTENTE: _____						

DATI INFRASTRUTTURA						
PARAMETRO	STANDARD*	N° POSTI LETTO				NOTE
		Presenti	Occupati	Liberi		
Capacità Alloggiativa interna	☐ Minimo					
	☐ Medio					
	☐ Ottimale					

* Standard Minimo: con camere non indipendenti, aventi un numero maggiore o uguale di 6 posti letto e servizi igienici in comune.

Standard Medio: con camere indipendenti aventi un numero minimo o uguale di 6 posti letto, aventi ciascuno un punto luce e una presa elettrica, con servizi igienici in

Standard Ottimale: con camere indipendenti da 2/4 posti letto, aventi ciascuno un punto luce e una presa elettrica, con servizio igienico annesso.

PARAMETRO	N. PERSONALE ALLOGGIATO ALL'INTERNO	PERCENTUALE PERSONALE ALLOGGIATO ALL'INTERNO SUL TOTALE	UOMINI	DONNE	NOTE
Tensione Abitativa interna		_____%	N. _____ ____%	N. _____ ____%	

PARAMETRO	N. POSTI DISPONIBILI	UTILIZZATI	LIBERI		NOTE
Parcheggi veicoli civili					

Fig. 7 Except from a "Technical Capacitive Record" developed by the Army General Staff Working Group and sent to the unit to share the needs with the stakeholders community

Fig. 8 Existing databases

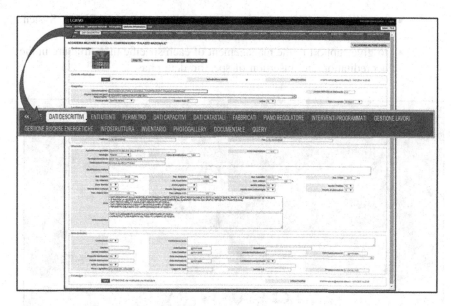

Fig. 9 Intermediate release of "LC2EVO-Infra" by product

- to illustrate the development which the LC2EVO System Infrastructural FAS has reached;
- to give support to the management of each specific infrastructure in the accreditation and insertion of specific data phases.

- of the FAS evolution and perfectioning following feedbacks coming from final users.
- This is the most peculiar and interesting phase of this case study (Figs. 10 and 11).

3   Conclusions

This case study is the F.A.S. pilot project and it is part of the LC2EVO "System of the Systems."

It was launched for the strategic management and the Army real estate property survey, both at a central and at a local level.

The infrastructure F.A.S. is the SW which we carried out over the last 8 months, accessible through the Web portal and not in a Stand-Alone modality. It completely meets the Army infrastructure requirements and it is therefore fully satisfactory.

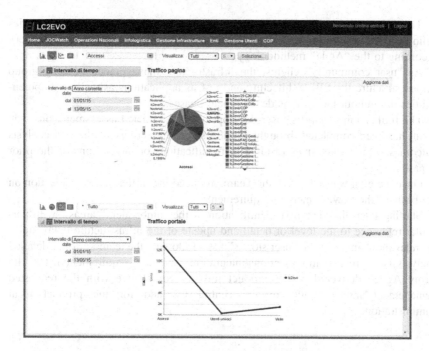

Fig. 10 "Statistics Function in LC2EVO"

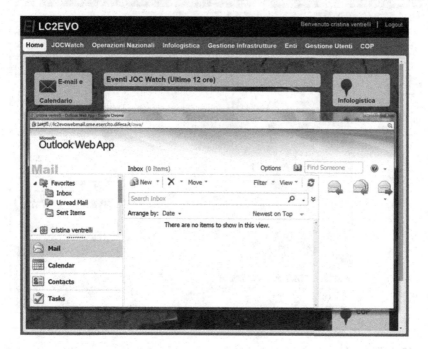

Fig. 11 "E-mail" function in LC2EVO

The SW was developed by a mixed team, who worked on the user stories definition and on the priority identification (in the various Product Backlog set), according to the "Agile" methodology.

As "user community" clients, the infrastructure management working group helped structure the SW architecture and test its applications as well as its potentialities throughout the various development phases.

The further joint analysis (user-developer) and the related assessment, once the sprint has been completed, brought about a screening activity of the first analysis outcomes, which sometimes resulted in modifications or integrations of the prior requests.

Once the goal was achieved, the Team produced the further priority selection on the basis of the newly emerged requirements.

Starting from the very first identification of the "ambivalent" areas—with particular reference to the level of detail and update of the infrastructure data—up to the relevant changes of the "user stories" list—following the team's further detailed analysis (with related modifications/integrations and new requirements), the "Ita Army Agile" delivered a SW Product fully in compliance with the requested management model, both for the central C2 and for the product local administration.

Fig. 12 LC2EVO-Infra: access from "Infologistica"

The SW was made visible and ready to use before being fully maximized in its functions, at any levels and "specific" fields. That confirmed its valuable cost-efficient features, its maximized functionalities, and implementation times, also from the consignee point of view.

Moreover, the role of the consignee was highly appreciated since it could provide feedback as well as the considerations of the Army General Staff even at an intermediate delivery phase. Hence, the F.A.S. system has become an essential support for the everyday work of the local users, also on the ground of the user friendly and intuitive interface.

The infrastructure F.A.S. is accessible by selecting the item "Infologics" or "Infrastructure management" on the "LC2EVO" panel, depending on the needs, respectively, for the C2 or for updating data (Figs. 12, 13 and 14).

The SW's continuous monitoring in the real environment allowed us to pinpoint the criticalities and the related improvement opportunities with the aim of devising further possible functions for the development of the "LC2EVO" system, based on the results achieved by the infrastructure subsystem.

The continuous use of the System will maximize the benefits for the whole influencers and stakeholders community thanks to the "Ita Army Agile" adaptive features and qualities which always allow to find the right solution to the Army's ever-changing and unpredictable requirements.

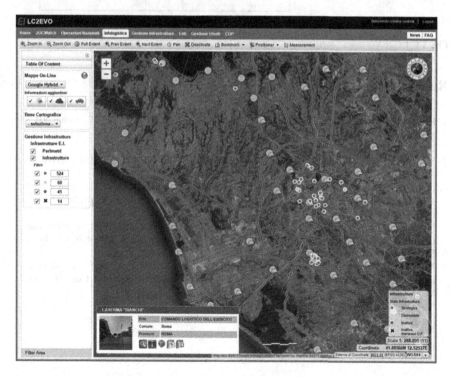

Fig. 13 LC2EVO-Infra: additional information in "Infologistica"

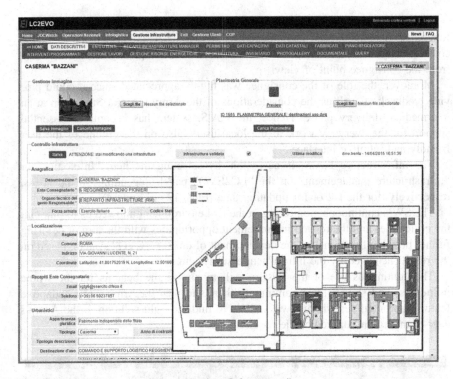

Fig. 14 LC2EVO-Infra: access from "Gestione Infrastrutture"

References

1. Ministero della Difesa, "Conferenza nazionale sulle infrastrutture militari", Roma 10–11 novembre 1986
2. Legge di stabilità 2014 – n° 147 del 27.12.2013
3. Stato Maggiore dell'Esercito, "Piano per la Revisione dello Strumento Militare Terrestre", 2013–2024
4. Cotugno F, Messina A (2014) Implementing SCRUM in the army general staff environment. In: The 3rd international conference in software engineering for defence applications, *SEDA* Roma, Italy, 22–23 Sept 2014
5. Messina A (2014) Adopting Agile methodology in mission critical software production. Consultation on cloud computing and software closing workshop, Bruxelles, 4 Nov 2014
6. Stato Maggiore Difesa, "Budget"; "GE.PA.D.D."
7. Messina A, Cotugno F (2014) "Adapting SCRUM to the Italian army: methods and (open) tools". In: The 10th international conference on open source systems, San Jose, Costa Rica, 6–9 May 2014. ISBN: 978-3-642-55127-7

Consumer Electronics Augmented Reality in Defense Applications

Cristian Coman, Pascal van Paridon and Franco Fiore

Abstract This paper presents an initiative to explore the potential of commercial electronics technologies, in particular augmented reality (AR), in the military domain. Although the ruggedization, reliability, and power consumption are challenges that prevent most of the hardware components from directly being used in operations, the consumer software components are confidently being used in the development process of operational military capabilities. The project has concluded that availability of advanced consumer electronics stimulates the development of new military concepts and supports rapid prototyping and software requirements elicitation.

1 Introduction

Wearable computing technologies have been advertised as a potential solution to enhance the human–computer interface (HCI) in case of mobile applications. Augmented reality (AR) is an emerging technology that has the potential to solve some of the HCI challenges in wearable applications, by presenting the information

---

DISCLAIMER: Any opinions expressed herein do not necessarily reflect the views of the NCI Agency, NATO and the NATO Nations but remain solely those of the author(s).

C. Coman (✉) · P. van Paridon · F. Fiore
NATO Communications and Information Agency, The Hague, Netherlands
e-mail: cristian.coman@ncia.nato.int

P. van Paridon
e-mail: pascal.vanparidon@ncia.nato.int

F. Fiore
e-mail: franco.fiore@ncia.nato.int

© Springer International Publishing Switzerland 2016
P. Ciancarini et al. (eds.), *Proceedings of 4th International Conference in Software Engineering for Defence Applications*, Advances in Intelligent Systems and Computing 422, DOI 10.1007/978-3-319-27896-4_5

into a contextual relevant view and by minimizing the user interaction. In many combat situations, soldiers cannot interact with the computing equipment and "hands-free" solutions developed in the context of AR technologies open new ways of addressing this problem.

During the past years, the North Atlantic Treaty Organization (NATO) Communications and Information (NCI) Agency has monitored the evolution of the AR technologies and identified a few use cases where such technologies can support the military wearable computing. The Italian Ministry of Defence is currently running soldier modernization programs (e.g., "Soldato Futuro" and its future development: Land Command and Control Evolution—LC2EVO), which focus on wearable computing as well [1]. The results presented in this paper are derived from the cooperation between Italy and NCI Agency in defining requirements for AR technologies and in rapid prototyping of AR solutions for soldier modernization.

NCI Agency has supported NATO and nations with the development and acquisition of information systems in various domains. Different software development approaches (waterfall, spiral, and agile) have been utilized by NCI Agency in the development of these capabilities. In acquisition projects, the waterfall methods are commonly employed at NCI Agency, but other solutions such as agile development were being introduced in the acquisition process [2]. Spiral and agile software developments have been the preferred approaches for research and development project such as Multi-sensor Aerospace-ground Joint Intelligence, Surveillance, and Reconnaissance Interoperability Coalition (MAJIIC) [3].

A variation of the agile software development methodology, which is based on availability of consumer electronics, has been employed in the project reported in this paper.

The rest of the paper is organized as follows. An overview of augment reality technologies available to commercial consumers is presented in Sect. 2. Operational considerations formulated in the beginning of the project and at subsequent review stages are presented in Sect. 3. The software design approach and some of the results of the project are presented in Sect. 4. Finally, Sect. 5 concludes the paper.

2 Consumer Augmented Reality

AR is not new to the military domain, and for many years, advanced fighter jets have been controlled with the aid of sophisticated AR helmets. Both the hardware and software AR components have been reengineered during the past years to open the market to commercial consumers.

AR devices are emerging into our day-to-day life, and projects such as Vuzix's M100, Epson's Moverio BT-200, Google's Glass (at the moment transitioning to a follow-up project), Sony's SmartEyeglass, or Microsoft's HoloLens have already established a growing consumer community (Fig. 1).

AR systems combine real and virtual information in real time by aligning (also called registration) real and virtual objects in the field of view of the user.

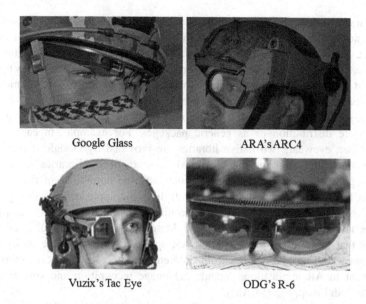

Google Glass   ARA's ARC4

Vuzix's Tac Eye   ODG's R-6

Fig. 1 Examples of commercial AR glasses considered in the military domain

Two common registration approaches for mixing the visual reality around a user with augmenting information are as follows:

- *Geospatial registration*, when the alignment criteria used are the geographic location (often collected from a global positioning system—GPS) and orientation data (commonly collected from a compass) and
- *Video registration*, based on video data collected with a camera integrated in the device and object recognition techniques.

The geospatial registration method is useful when the range to the location of augmenting information is large (tens or hundreds of meters—commonly the human visual range). In case of geospatial registration augmentations, it is important to have a database of geolocated information which is relevant to the area where the dismounted operation is conducted. For example, such databases are commonly maintained in the military domain to indicate vulnerable points or hot spots in the joint area of operations (JOAs).

Video registration relies on image processing techniques, which are used to automatically detect objects of interest in the vicinity of the user and augment the visual perception of these objects with drawings and labels. A simple example of such an application in the defense domain is the automatic recognition and labeling of unexploded ordinance encountered in dismounted missions.

The availability of commercial AR hardware components has stimulated the development of the software packages that incorporate advanced video processing functions, which are commonly employed in AR applications (e.g., image segmentation, image recognition, and image labeling). The availability of AR eyewear to general consumers also expanded the domain of utilization of these technologies to other non-traditional AR domains such as automated maintenance instruction assistance (e.g., automotive industry), virtual expert teams, and remote medical assistance. On the military side, the land forces were offered AR capabilities mainly to support dismounted operations [1]. However, the adoption of AR solutions in combat operations is still in its early phase.

On the software side, specialized AR components are either provided as part of the hardware distribution or as generic packages. For example, in case of the Google Glass eyewear, two native libraries are provided by Google: the Google Development Kit (GDK) and the Mirror API. These specific libraries are used to develop optimized Google Glass applications, and the portability of these applications is limited. If the possibility to install the same application on AR glasses provided by different vendors is a key requirement, more generic development packages are available. Wikitude [4] and Metaio [5] are two popular software packages that allow developers to build AR applications that can be easily used with eyewear and mobile devices from different vendors. Key functionalities that are relevant to AR applications include 2D image recognition and simultaneous localization and mapping (SLAM).

Selection of AR eyewear remains a difficult task, with vendors advertising many features of their technologies, which are not always relevant to defense requirements. Another sensitive area in the selection process is the current availability of products versus availability in the future. The volatility of the AR consumer technology market is illustrated by recent decision of one of the leading companies, Google, to terminate their Google Glass program and look into a different technological solution (see http://www.magicleap.com/). In a report published by Augmented Reality Organization in 2014, the maturity level of major AR eyewear producers shows that most of the initiatives are at the exploratory stages where limited orders are accepted from AR developers' enthusiasts [6].

3 Operational Considerations

The Italian Army General Staff Logistic Department in cooperation with the NCI Agency has established in 2014 the project EYE CATCH, which aims at investigating the feasibility of AR in dismounted operations. The project addresses requirements elicitations through fast prototyping and has adopted agile software development practice and utilization of commercially available AR devices.

During the initial phase, the Vuzix M100 and the Google Glasses have been considered to refine the use cases relevant to the soldier modernization program. The project has also considered common NATO information repositories as the source of augmented information. Most of these repositories contain geolocated reports recorded either through the command and control (C2) channels or though the intelligence, surveillance, and reconnaissance exploitation.

Fig. 2 Display of information about past events in the proximity of dismounted soldier through AR notifications in the user's field of view

The use cases considered within the project cover presentation of information about past events in the area of operation, coordination between the members of a squad during a mission, and notifications through messaging and reporting.

For example, in many dismounted operations, it is important to identify hot spots where critical events, such as improvised explosive device (IED) strikes, happened in the past. JOCWatch is an example of a repository where NATO maintains information about events on the battlefield. Figure 2 illustrates how the information recorded in JOCWatch is presented to the dismounted soldier through an AR solution developed within the Eye Catch project. The events are shown in the user's field of view using labels overlaid on the video image collected by the AR glass. The dots in the radar display in the top left corner in Fig. 2 represent events available in the JOCWatch database, which are located within a certain range around the user. The highlighted sector on the radar display is an estimation of the orientation of the user's field of view. For events positioned in the center of the field of view, the user has an option to select them by a simple touch of the glass frame and display additional.

4 Software Engineering Approach

Acquisition and prototyping are two common approaches for developing information systems in NATO. The waterfall software engineering method is common for acquisition projects. The process is commonly implemented through the following steps:

- Minimum military requirements (MMRs),
- System requirements specification,

- Critical design review,
- Implementation,
- Test and acceptance, and
- Operation and maintenance.

Changes of requirements are normally implemented through the Engineering Change Request process, which often leads to end user dissatisfaction, delays, and additional costs in the project. To circumvent such inconveniences, agile methods have been considered in some of the NATO projects into address the development of partial components [2].

Spiral development was at the core of the MAJIIC project, a research and development initiative of nine NATO nations looking at improving the interoperability within the Joint Intelligence, Surveillance and Reconnaissance (JISR) domain. The MAJIIC approach is based on cycles of about one year during which military requirements were transformed in technical requirements and implemented in information systems [3]. One technical and one operational test phases are used during each of this cycle to verify compliance to requirements and update the requirements. More recently, the continuation of the project called MAJIIC 2 has considered software-oriented architecture (SOA) through the spiral development approach.

The Eye Catch project concentrates on concept development through use of consumer AR electronics. The importance of user adoption and the development of operational concepts are considered critical in all new capability development initiatives. These are log processes, and early initiation through the use of consumer electronics can significantly reduce the risks of delivering a "fit-for-purpose" military AR capability.

From the beginning, an agile development method was adopted within the project [7]. The agile approach had to be adjusted to be able to make use of commercially available components.

A schematic of the agile software (SW) life cycle adopted in the project is depicted in Fig. 3. The minimum military requirements are captured in a less formal manner through the discussion with the customer. These requirements are then revised and expressed through user stories recorded in the project backlog. At this stage, the detail technical requirements are still not clearly articulated, but the key behaviors of the functionality are captured in these stories. User priority, time, and effort estimates are used to rank the stories and select the ones that will be implemented through the sprints of the project.

AR electronics market is rapidly evolving at the moment, and new eyewear components are released at very short interval. The design process is based on the user stories, and the same time considers the availability of the AR hardware.

The development process is closely integrated with the testing activities. At this phase, frequent interaction with the end user is required in order to understand the operational condition and refine the requirements and the implementation.

Fig. 3 An iteration of the agile software life cycle influenced by the availability of AR eyewear and AR development frameworks

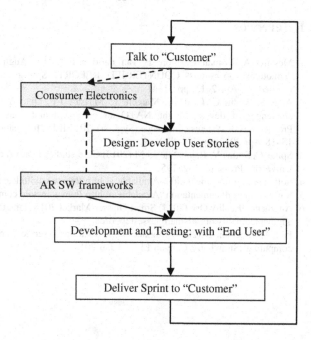

AR software development frameworks are made released to the consumer market at a high frequency. These frameworks significantly reduce the development time, and they have been considered in the Eye Catch project.

Despite the flexibility in the requirements definition process, a formal agreement is still established between the customer and the developer, which includes clear and measurable deliverables, mainly expressed in terms of level of effort.

5 Conclusions

The use of AR consumer electronics in the development of military capability was addressed through the project presented in this paper. The approach demonstrated that the consumer electronics can be used for concept development through rapid prototyping, which reduces the risk of deploying AR capabilities into operations.

A modified agile software development methodology was adopted in the project in order to adapt to a dynamic AR consumer electronics market, which is characterized by frequent releases of new components and implicitly a short life cycle of the hardware components (of about one year).

References

1. Messina A, Coman C, Fiore F, van Paridon P (2015) Augmented reality in support of dismounted operations. COUNTER-IED REPORT, Spring 2015, Delta Business Media Limited, 17 Apr 2015, pp 31–41
2. Aker S, Audin C, Lindy E, Marcelli L, Massart J-P, Okur Y (2013) Lessons learned and challenges of developing the NATO air command and control information services. In: Proceedings of systems conference (SysCon), 2013 IEEE International, IEEE, Orlando, Florida, 15–18 Apr 2013, pp 791–800
3. Spitas C, Spitas V, Rajabalinejad M (2013) Case studies in advanced engineering design. Delft University Press, pp 182–195
4. Software and documentation. Available at http://www.wikitude.com/
5. Software and documentation. Available at http://www.metaio.com/
6. AugmentedReality.Org (2014) Smart Glasses Market 2015. Jan 2014. Available at http://www.augmentedreality.org/#!smartglassesreport/c88h
7. Fox A, Patterson D (2014) Engineering software as a service: an agile approach using cloud computing. Strawberry Canyon LLC, 3 Jan 2014

Agile: The Human Factors as the Weakest Link in the Chain

Ercole Colonese

Abstract Despite the excellent results achieved by the agile methodologies, software projects continue to fail. Organizations are struggling to adopt such methods. Resistance to change is strong. The reasons, related to the culture and people, are many. The human factor is the weakest link in the organizational chain. Many inhibitors prevent the adoption of good practices. C.G. Jung stated in 1921 "how difficult it was for people to accept a point of view other than their own." Based on his mental process, two American researchers, mother and daughter, have created the homonymous model Myers–Briggs Type Indicator (MBTI). The tool helps us to better understanding others and ourselves: how to gather information from the outside, how to elaborate information and make decisions, and how to act afterward. MBTI supports Agile in creating successful teams: better communication, share of leadership, effective problem solving, stress management, etc. Psychological Types at Work, 2013 provides a guide to these items.

1 The Contest Today

1.1 Chaos Report 2014

The Chaos Report 2013 results by Standish Group depicts an increase in the rate of success projects, with 39 % of all projects succeeding (deliver on time, on budget, with required features and functions); 43 % were challenged (late, over budget, and/or with less required features and functions); and 18 % failed (canceled prior to completion or delivered and never used) [1].

Ercole Colonese: Member of Technical and Scientific Committee.

E. Colonese (✉)
TUV SUD Academy - Technical and Scientific Committee, Milano, Italy
e-mail: ercole@colonese.it

© Springer International Publishing Switzerland 2016   59
P. Ciancarini et al. (eds.), *Proceedings of 4th International Conference
in Software Engineering for Defence Applications*, Advances in Intelligent
Systems and Computing 422, DOI 10.1007/978-3-319-27896-4_6

The increase in success depends on many factors, such as methods, skills, costs, tools, decisions, optimization, internal and external influence, and team chemistry. Understanding the importance of skills needed to be an effective project sponsor (i.e., Product Manager) as well as improving the education of project responsible as a project management profession (i.e., ScrumMaster) can be directly linked to success rate. Smaller projects and Agile projects represent a big portion of success ones. However, the success gained with a relevant cost: project overhead and reduction in value and innovation. Often, the organizations adopt the agile methodology as "the last resort" within projects in trouble.

Project health checks, retrospective sessions, dashboards, and tracking systems provide an early project status evaluation and warning, so corrective actions can be taken appropriately and timely. The most parts of organizations perform a sort of postmortem review, but very few of them register information, analyze data, gather trends, and improve processes. Many times the information is lost or forgotten.

Agile methodologies propose alternatives to classic software development approach. Positive results from the projects adopting these methods encourage us to think that it is possible to develop software on time, on budget, and with required functionalities and quality. The previously mentioned report by Standish Group confirms this trend reporting a percentage of small projects successfully completed (76 %) significantly higher than the percentage of large ones (10 %). "Think Big, Act Small" is the title of the report. The use of agile processes is in the report as one of the ten factors that determine the success, as shown in Table 1.

The report gives following definitions for the ten success factors:

Executive Management Support: The executive sponsor (Product Owner in Scrum approach) is the most important person in the project. He/she is ultimately responsible for the success and the failure of the project. The factor scores 20 points in the small project.

User Involvement: The research clearly shows that projects that lack user involvement perform poorly. User participation has a major effect on project resolution both on large and on small one. The factor scores 15 points.

Table 1 Success factor and points

Success factor	Points
Executive management support	20
User involvement	15
Optimization	15
Skilled resources	13
Project management expertise	12
Agile process	*10*
Clear business objectives	6
Emotional maturity	5
Execution	3
Tools and infrastructure	1

Optimization: Is very important in small projects, especially if the optimization is a project with small labor content and fast delivery. The factor scores 15 points.

Skilled Resources: Skilled resources require more respect. People make a project, and success is on their shoulders. This is especially true in small projects. The success factor scores 13 points.

Project Management Expertise: It is essential to controlling the project status. In particular in small projects, the control is a continuous interaction and feedback with/from the development team to make effective and quickly decisions. The factor scores 12 points.

Agile Process: Embodies the small project philosophy. The agile process directly addresses user involvement, executive support, and the other success factors. The agile process scores 10 points for small project.

Clear Business Objectives: Is less important in small projects than in larger ones. Still, small projects need clear business objectives. Customer representative (i.e., Product Owner in Scrum methodology) actively participates in the project and reduces the needs for predefined and stable objectives. The success factor scores 6 points.

Emotional Maturity: Covers the emotional state of the project environment. Projects get resolved and reach success within a healthy ecosystem. Success factor scores 5 points.

Execution: Is the process that governs and controls the project. Generally, controls focus on financial and procedural aspects. It is important but not decisive for the success of small projects. The success factor scores 3 points.

Tools and Infrastructure: They can help project succeed, but are not relevant (necessary condition but not sufficient). Organization must be very careful not to rely too much on tools for the project's success. The success factor scores only one point.

The first five success factors focus on the execution of small project skills and provide the greatest benefit for the success. The first three success factors account for 50 % of the points. All first five account for three-quarters of the points. The last five success factors help, but provide the least benefit for success, with a total of 25 points out of 100.

One interesting perspective of the reported elements is the *Skilled Resources* and the *Emotional Maturity* of the organization. The scores are, respectively, 13 and 5 points, and the influence on the success, in my opinion, is very relevant as presented in this article.

Agile methodology seems to be sufficient to solve the "software crisis" challenges: the inadequate culture of the organization, the difficulty to break down projects into smaller sized initiatives, the lack of active involvement of the business into the project, and the technical and managerial personnel not adequate to interpret the roles assigned.

Results demonstrate how hard is for the organizations to change paradigm and adopt agile methodologies. Processes, method, and techniques are well understood and tools installed and utilized, but may projects continue to fail. Why?

1.2   Boehm and Turner Study

Barry Boehm and Richard Turner conducted a study on characteristics of agile and plan-driven methods to provide guidance in balancing the agility and discipline required for successful software development [2]. One of the most significant results of their analysis was the realization that while methodologies, management techniques, and technical approaches are valuable, the most critical success factors are much more likely to be in the area of people factors.

In their article, Boehm and Turner discussed five areas where they believe significant progress we can make *staffing*, *culture*, *value*, *communications*, and *expectations* management [3]. Following is reported a brief summary of five areas that I will complete with psychological elements in this article.

1.2.1   Staffing

"In essence—they say—software engineering is made of people, by people, for people." The two primary categories of players in the software development world are customer and developers. Many problems in project failure belong to this area. Customer and developers are not fully able to collaborate to reaching the common goal: deliver success software in time, within budget, and with required features. Why? They are not able to share goals and risks. The final objective is common to both sides—customer and development—but they do not share it. Each side assumes that the other one is responsible for failure. The two parts do not match, and the result is unsuccessful. I will give my personal though on this issue.

Customer. "Unfortunately—they say—software engineering is still struggling with a separation-of-concerns legacy that contends translating customer requirements into code is so hard that it must be accomplished in isolation from people concerns, including customers." In agile projects, the customer representative (i.e., Product Owner in Scrum method) is required to have some characteristics that are not always present in such a people. Boehm and Turner identified required skills as following: *collaboration*, *representativeness*, *authorization*, *commitment*, and *knowledge* (CRACK). During their research in California, they discovered that some projects go well at the beginning, failed during the execution because of customer representatives changed during the execution, and the new ones did not have the right skills as the previous have. The writers did not specify which competencies the representatives missed. In any case, the human factor resulted crucial to the project unsuccessful.

Developers. Critical people factors for developers using agile methods include *amicability*, *talent*, *skill*, and *communication* [4]. Alistair Cockburn addressed levels of skill and understanding required for performing various method-related functions, such as using, tailoring, adapting, or revising a method. He has identified a three level of software development method based on the martial art Aikido [5].

1.2.2 Culture

Culture is the second area of people critical factor. In agile environment, people feel comfortable and are empowered when they can work with a level of freedom to defining the work and the address problems. This is the typical environment, where each person is expected and trusted to do whatever necessary to complete the necessaries tasks for project success. Unfortunately, trusting, empowerment, and delegation are not enough developed in the organizations to declare complete transformation from traditional approach to the new agile paradigm. Some changes yet need.

1.2.3 Value

Values come from people. Different people have different values. One significant challenge is to reconcile different value propositions made on software by customers, users, developers, and other stakeholders. Unfortunately, all requirements, use case, and user stories have the same importance and discussions arise on doing all or none at all. Defining value requires people to understand value to the business, prioritize requirements, and trust their development. Again, it is a cultural, attitude, and organizational issue.

1.2.4 Communications

Both customer and developers have a different view of the problem. Customer lives the "problem domain" and development the "solution domain." Not easy reconciliation between two perspectives. Customer representative and development speak two different languages. They can say the same thing in two different forms and continue to discuss many hours without reaching an agreement. Both they know the communication techniques and use them, but they continue to do not agree. Psychological types described in Sect. 2 will clarify this concept and suggest a possible approach to problem solution.

1.3 The Question on the Table

There is still the open question: Why is it so difficult for the organizations to adopting the agile paradigm, and why projects continue to fail?

The human factor is the weakest agile link of the chain! It is easier to change processes, technologies, methods and techniques that do not people!

Which organizations are able to produce such a radical change of their own culture? What constraints and resistance they must overcome to push the business and production, customers and suppliers, working together to achieve common goals?

Each project represents a major investment for the client. To its success are tied business objectives, organizational change, innovation, competitiveness, and market challenges. Yet, even today, too many projects do not achieve their goals, or they achieve them only partially. The analysis attributes the failures to inadequate management of the requirements and changes, poor support from management, inaccurate estimates and unrealistic schedules, inadequate or lack of risk management, ineffective communication. What these items have in common? The human factor!

The methodologies abound as well as maturity models. Agile has defined new ones. The baton passes, once again, in the hands of the interpreters: people!

The representative of the customer, the project leader, and the team, i.e., everyone, involved in the project with the common goal to successfully completing the project are required to have skills and attitudes such as collaboration, representativeness, authority, responsibility, business knowledge, and expertise on technical and methodological aspects.

Why is it so hard to find a group with all of these competencies combined?

2 A First Response

Carl Gustav Jung[1] provided a scientific response already in 1921 by publishing *Psychological Types* [6]. His theory explains some of the apparently random differences in people's behavior. The result of a two-decade research on his patients and others, Jung affirmed that despite the multiple diversity and uniqueness of each person, all use the same mental processes to perceive reality, to process information, and to decide how to act. Jung found predictable and differing patterns of normal behavior. His theory of psychological types recognizes the existence of these patterns, or types, and provides an explanation of how types develop.

The first process is to gathering information from the outside world (*Perceiving*) through the senses (*Sensing*) or intuitively (*Intuition*). The second process is to elaborating (*Judging*) that information and coming to conclusion in a logical and rational manner (*Thinking*) or by use of sensitivity (*Feeling*) (Fig. 1).

Everyone uses these four essentials processes daily in both the external world and the internal world. Jung called the external world of people, things, and experience *Extraversion* and the internal world of inner processes and reflections *Introversion*. These four basic processes used in our both external and internal worlds give us eight different ways of using our mind.

Jung believed everyone has a natural preference for using one kind of Perceiving process (Sensing or Intuition) and one kind of Judging process (Thinking or Feeling). He also observed that a person is drawn toward either external or internal

---

[1]C.G. Jung (1875–1961) was a Swiss psychologist and psychotherapist who founded analytical psychology.

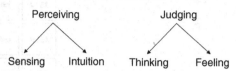

Fig. 1 Jung's theory mental process

world more than the other. As we exercise our preferences, we can develop distinct perspective and approaches to life and human interaction.

It seems to glimpse a possible solution to our problem:

Why different persons have so different approach and a different view of the same problem?

The variation in what we prefer, use, and develop leads to fundamental differences between people. The resulting predictable patterns of behavior form psychological type.

2.1 MBIT Model

Two American researchers, mother and daughter,[2] building the eponymous model Myers–Briggs Type Indicator (MBTI), further elaborated the dichotomous model of Jung.

The MBTI is a self-reported questionnaire designed to make Jung's theory of psychological types understandable and useful in everyday life. Its results describe valuable differences between normal, healthy people—differences that can be the source of much misunderstanding and miscommunication.

The MBTI will help us to identify our strengths and unique gifts. We can use information to better understanding ourselves, our motivations, our strengths, and potential areas for grow. It will also help us to better understand and appreciate those who differ from us.

Understanding MBTI type is self-affirming and enhances cooperation and productivity.

The MBTI is used in many environments and with different objectives: self-development, communication, leadership, team building, problem solving, career development and exploration, relationship counseling, academic counseling, organization development, education and curriculum development, and diversity and multicultural training.

Myers–Briggs added to the eight Jung's types schema another dichotomy (Judging–Perceiving) building a model with 16 different types as shown in Fig. 2.

---

[2]MBTI's authors: Katharine Cook Briggs (1875–1968) and her daughter, Isabel Briggs Myers (1897–1980).

Fig. 2 MBTI representation
of types

ISTJ	ISFJ	INFJ	INTJ
ISTP	ISFP	INFP	iNTP
ESTP	ESFP	ENFP	ENTP
ESTJ	ESFJ	ENFJ	ENTJ

The result of the MBTI test will position each person in a specific cell of the model depending on the expressed preferences. It is not matter of this article to explain the model. What I would like to present is the helpful usage of the tool (and of the theory).

I am not a psychologist; I am a software engineer, a management consultant, and a project manager. What I present is the mere result of the MBTI application in my work and the benefits collected. Many success projects completed on time, within budget, with required quality, and with satisfaction of both client and development team. All those projects and consultancy engagements have been managed applying the competencies built on the application of MBTI guidelines [12].

2.2 What Can We Learn with MBTI?

Looking in depth at each type, we can better understand how people in the team gather information, analysis data, make decisions, and act within the team [7–9].

2.2.1 Energy

The *Extraverted* (*E* in the model) person directs and receives energy from the outside world. He/she prefers action over reflection, talks things over in order to understand them, prefers oral communication, shares his/her thoughts freely, acts and responds quickly, extends himself/herself into the environment, and enjoys working in groups.

Vice versa, the *Introverted* (*I* in the model) person directs and receives energy from the inner world. He/she has opposite preferences of Extraverted: prefers reflection over action, thinks things through in order to understand them, prefers written communication, guards his/her thoughts until they are (almost) perfect, reflects and thinks deeply, defends himself/herself against external demands, enjoys working alone or with one or two others.

2.2.2 Information Gathering

The *Sensing* (*S* in the model) person prefers to gather information in a precise and exact manner. He/she likes specific examples, prefers following the agenda, emphasizes the pragmatic, seeks predictability, sees difficulties as problems that need specific actions, focuses on immediate applications of a situation, and wants to know what is (He/she is a "Practical person").

The *Intuition* (*N* in the model) person, as the opposite of Sensing, prefers to gather information in a novel or inspired manner. He/she likes general concepts, departs from the agenda if necessary, emphasizes the theoretical, desires change, sees difficulties as opportunities for further exploration, focuses on future possibilities of a situation, and wants to know what *could be* (He/she is an "Intuitive person").

2.2.3 Decision Making

The *Thinking* (*T* in the model) person seeks general truths and objectivity when making decisions. He/she questions first, knows when reason is needed, wants things to be logical, has a cool and impersonal behavior, remains detached when making decisions, controls the expression of his/her feeling, and overlooks people in favors of tasks (He/she is a "Logical person").

The *Feeling* (*F* in the model) person, as opposite of Thinking, seeks individual and interpersonal harmony when making decisions. He/she accepts first, knows when support is needed, wants things to be pleasant, has a warm and personal behavior, remains personally involved when making decisions, expresses his/her feelings with enthusiasm, and overlooks tasks in favor of people (He/she is a "Sensible person").

2.2.4 Lifestyle

The *Judging* (*J* in the model) person likes to come to closure and act on decision. He/she likes things to be settled and ordered, finishes tasks *before* the deadline, focuses on goals, results, and achievements, establishes deadlines, prefers no surprises, prefers to be conclusive, and commits to plans or decisions (He/she is a "Precise person").

The *Perceiving* (*P* in the model) person, as opposite to Judging, prefers to remain open and adapt to new information. He/she likes things to be flexible and open, finishes tasks *at* the deadline, focuses on processes, options, and openings, dislikes deadlines, enjoys surprises, prefers to be tentative, and reserves the right to change plans or decisions (He/she is "Last minute person").

Table 2 MBTI preferences summary

Preference for	Affects a person's choice
EI Extraversion or introversion	To focus the dominant process on the outer world or on the world of ideas
SN Sensing or intuition	To use one kind of perception instead of the other when either could be used
TF Thinking or feeling	To use one kind of judgement instead of the other when either could be used
JP Judgement or perceiving	To use the judgement or the perceptive attitude for dealing with the outer world

2.2.5 Summary of the Four Preferences

The MBTI authors describe the value of the model as summarized in Table 2. They also specify:

> Personality is structured by four preferences concerning the use of perception and judgment. Each of these preferences is a fork in the road of human development and determines which of two contrasting forms of excellence a person will pursue [9].

3 Why Psychology in Software Development?

What does psychology with agile development? It does [13]. In addition, it provides a possible answer to the difficulty of finding a group of persons and building a team with all the required characteristics and attitudes to successes. Jung stated that all of us, in a more or less accentuated level, hold the characteristics mentioned in the previous sections. Understanding different types can help building teams, managing effectively communications and critical situations, solving problems, and managing stress.

MBTI, in particular, can help sponsor, project manager, and member of teams to operate in an effective and productive way. Once we have determined our four preferences, we will have a four-letter type (such ENTJ in my personal case). Particularly important is the notion of dynamic relationship. For each type, one preference will be developed more than the others will. This is called the "dominant" function and reflects our contribution to the world. If the dominant function is one of the Gathering Information (either S or N), then the second function, called the "auxiliary" function, will be one of the Decision-Making preferences (either T or F) and vice versa.

The third and fourth functions develop subsequently and differ in the use and confidence in them. The fourth function, also called "Achilles heel," is the last one and represents the area most vulnerable of the person.

Each has his/her preferences that become strengths or weaknesses depending on the specific conditions. It is not normal, however, to think that this is a limit, a negative element. Each one gives the best of himself/herself when he/she gives the opportunity to express his/her "dominant" preference. Ignoring that others can have preferences different from ours, and they can think and act differently from the way we want, it means not being able to grasp the best that they can give to the group. Negatively evaluating others because they do not act as we are asking is an act of ignorance and managerial myopia at the same time. Is to miss the best of each!

3.1 MBTI Within Agile Projects

An agile team needs to effectively communicate, create an effective team, share leadership, solve problems effectively, and manage stress and critical situations.

Three roles are very important in an agile project: customer representative (i.e., Product Owner in Scrum approach), project responsible (i.e., ScrumMaster), and development team. Each one has specific responsibilities and authorities and is required to possess specific competencies and attitudes.

According to Boehm and Turner [3], the Product Owner plays the most critical role. His/her responsibilities require characteristics and skills such as *collaboration*, *representativeness*, *authority*, *commitment*, and *knowledge* (CRACK). The continuous relationship with other roles and the need for communicating, negotiating, and collaborating require the client representative to possess attitudes that will be expressed more or less effectively depending on the psychological type. More details are given in Sect. 3.3.1.

ScrumMaster, as a project manager, plays another important role in agile projects. He/she is the primary point of contact for the development organization. His/her role gives the ScrumMaster-specific responsibilities: coach the team, servant leader, process authority, interference shield, impediment remover, and change agent. These responsibilities require characteristics and skills such as *knowledgeable*, *questioning*, *patient*, *collaborative*, *protective*, and *transparent* [14]. He/she can learn some characteristics, other are natural attitudes depending as the psychological type each one belongs to. More details are given in Sects. 3.3.2 and 3.3.3 [14].

The development team plays the third crucial role in agile projects. Its principal responsibilities include (in a Scrum terminology): perform sprint execution, inspect and adapt each day, groom the product backlog, plan the sprint, and inspect and adapt the product and the process (Sprint review and Sprint retrospective). These responsibilities require each team member having specific characteristics, skills, and attitude such as *self-organizing, cross-functionally diverse and sufficient, t-shaped skills, musketeer attitude, high-bandwidth communications, transparent communication, right-sized, focused and committed, working at a sustainable pace*, and *long-lived* [14].

3.2 How MBTI Can Benefit Teams

When team members understand their styles and those of others, they become more effective. A team member can use psychological type's preferences to better understanding himself/herself and how members relate to others. He/she can also understand contributions other team member can give to the project.

The MBTI specifically supports team members by: reducing unproductive work, identifying areas of strengths and possible areas of weakness, clarifying team behavior, helping to match specific tasks assignments with team members according to their MBTI preferences, supporting team members to understanding and better handling conflicts and stress situations, providing team members with perspectives and methods to successful solving problems, and maximizing the team's diversity in order to reach more useful solutions.

According to an article by Mary McCaulley, certain considerations allow to build teams that are more effective by considering psychological types.

> The more similar the types on a team, the sooner the team members will understand each other; the more different the types, the slower the understanding. Groups with high similarity will reach quicker decisions but are more likely to make errors due to inadequate representation of all viewpoints; groups with many different types will reach decisions more slowly (and painfully) but may reach better decisions because more viewpoints are covered. Leadership roles may shift as the tasks to be done require the skills of different types on the team. Team members who are opposite on all four preferences may have special problems in achieving understanding; members who share two preferences from each of the opposites may be "translator or facilitator". The person who is the only representative of a preference (e.g., the only Introverted) may be seen as a "different" from the other team members. Teams that appreciate and use different types may experience less conflict. Teams that are "one-sided" (i.e., have few different types) will succeed if they use different types outside the team as resources or if they make the effort to use their own less-developed functions as required by the tasks. One-sided teams may fail if they overlook aspects of problems that other types would have pointed out or if they stay "rigidly true to type" and fail to use other resources. [15]

McCaulley concluded:

> Good decisions will be made when basic facts and realities have been addressed (*Sensing*), when new possibilities have been discovered (*Intuition*), when unforeseen inconsistencies and consequences have been exposed (*Thinking*), and when important values have been protected (*Feeling*). [15]

3.3 Specific MBTI Benefits

To better understand how people communicate, decide, and act, MBTI suggests furthering dividing the four-letter type into various two-letter combinations. We can use these combinations as "lens" through which to viewing the interaction of people in a team.

3.3.1 Communicating Effectively

To reduce the unnecessary documentation and reduce bureaucracy sterile or harmful, team members must communicate better. People prefer to communicate in ways distinct (listen, understand, respond, interact). MBTI helps us to communicate better; it allows us to know how others interpret our messages and helps us to correctly interpreting their messages and feedbacks [11].

If the team is working on communication issues, the "Function Lens" can provide important insights. For example [8]:

STs prefer to use proven methods of communication.
SFs like to share their experience to help others.
NFs prefer to communicate in creative ways.
NTs like to debate challenging questions.

3.3.2 Building an Effective Team

An agile team makes the collaboration, the understanding, and the mutual respect its main strength. It is multidisciplinary and needs to be cohesive, result oriented, respectful of the diversity of each individual, able to facilitate collaboration and to appreciate the contribution of each member. MBTI helps us better understanding how they think and act: people process acquired information in a preferred manner by the senses (Sensing) or intuitively (Intuition). The team needs both skills [11].

If the team is dealing with cultural and change issues, the "Quadrants Lens" is useful. For example [8]:

ISs want to be careful and mindful of details.
ESs want to see and discuss the practical results.
INs want to work with ideas and concepts.
ENs want to maximize variety.

3.3.3 Sharing Leadership

An agile team takes important decisions at any time of the day (it assigns priorities, makes estimates, takes commitment, makes choices, and solves problems). MBTI helps us make better decisions: People make decisions based on logic and rationality (Thinking), or taking into account the impact on people (Feeling). The team needs both aspects: rationality and sensibility. An agile team needs facilitators, as well as leaders [11].

If the team is working with leadership issues, the "Temperament Lens" can help. For example [8]:

Fig. 3 Z problem-solving
model [10]

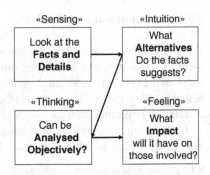

SJs value responsibility and loyalty.
SPs value cleverness and timeliness.
NFs value inspiration and a personal approach.
NTs value ingenuity and logic.

3.3.4 Effective Problem Solving

An agile team will solve problems at any time; most of the problems are critical and
complex; it needs the help of the most, according to the characteristics of the
individual members as required by the phases of the problem-solving process.
MBTI helps us to better solving problems: People solve problems using different
mind-sets and responsive to their psychological preferences; the problem-solving
process requires a sequence of skills that no one possesses entirely; people with
different psychological types can better solve complex problems [10, 11].

"Dynamic Lens" is particularly useful when teams are working on
problem-solving, decision-making, and stress-related issues. Figure 3 shows the
order in which we should correctly manage problems. Unfortunately, nobody has
preferences in this order. More types that are different need to correctly solving
problems [10].

3.3.5 Managing Stress and Critical Situations

Even an agile team can leave moments of tension, misunderstandings, and difficulties. MBTI helps us to better respond to stress: Everyone reacts differently to the
difficulties driven by the weaker preferences discovering its Achilles heel. We must
help the person under stress for the common good.

4 Conclusion

Agile methodologies, in conclusion, can afford to break down bureaucracies, reduce documentation, naturally accept changes during construction, be oriented to produce the final solution rather than follow complex and often unrealistic plans if people will learn about the others, respect the different preferences, and act capitalizing best suited mode to guide their thinking and actions!

Because the MBTI is a powerful tool, it is important that we use the ethical guidelines directly coming from the authors [8, 9].

References

1. The Standish Group (2013) Chaos Manifesto 2013, think big, act small, pp 1, 3, 4 https://larlet. fr/static/david/stream/ChaosManifesto2013pdf
2. Boehm B, Turner R (2004) Balancing agility and discipline: a guide for the perplexed. Addison-Wesley, Boston
3. Boehm B, Turner R (2003) People factors in software management: lessons from comparing agile and plan-driven methods. Cross Talk J Defense Software Eng
4. Highsmith J, Cockburn A (2001) Agile software development: the business of innovation. Computer, 34(9): 120–122, IEEE Computer Society Press Los Alamitos, CA, USA
5. Cockburn A (2006) Agile software development. Addison-Wesley, Boston
6. Jung CG (2011) "Tipi psicologici", Bollati Boringhieri, Torino, pp. 7–11, 359–446, 449–457, 475, 482–484
7. Briggs Myers I Revised by Kirby LK, Myers KD (1993) Introduction to type, 5th edn. CPP, Consulting Psychologists Press, Inc. Palo Alto, California
8. Hirsh SK (1992) MBTI: team building program. CPP, Consulting Psychologists Press, Inc, Palo Alto
9. Briggs Myers I, with Myers PB (1980, 1995) Gifts differing. Understanding personality type. CPP, Mountain View, California, pp 1–15, 115–122, 167–172
10. Kroeger O, with Thuesen JM, Rutledge H (2002) Type talk at work. DTP, Palo Alto, CA, pp 63–192
11. Colonese E (2013) Tipi psicologici al lavoro. Roma, Edizioni Nuova Cultura, pp 91–180
12. Colonese E (2010) Collaborazione nei progetti software: Quando clienti e fornitori collaborano per il successo dei progetti, *Qualità On Line, N.2–2009*, Aicq
13. Colonese E (2010) Psicologia ed economicità dei test, *Qualità On Line, N.3-2010*, Aicq
14. Rubin KS (2013) Essential scrum. A practical guide to the most popular Agile process. Addison-Wesley, pp 163–211
15. McCaulley M (1975) How individual differences affect health care teams. Health Team News 1(8):1–4

Rapid Prototyping

Nazario Tancredi, Stefano Alunni and Piergiuseppe Bruno

Abstract The developed methodology aims to study and define a platform for "Rapid Prototyping," in a full "Model-Based Design" approach, providing a significant reduction in the development time. The described approach allows us to test algorithm models, realized by the designer in Simulink™, directly on a real platform (prototype On Board Computer), in real time, to evaluate the performance in terms of computational load. The main advantages of this innovative approach with respect to the traditional one are an increased robustness of the algorithms before the integration (reduction of development time) and the ability to quickly evaluate different design solutions in terms of performance and reliability. With the implementation of the described design logic is possible to put the On Board Computer in the loop simulation starting from the early stages of development (CIL—computers in the loop). The proposed methodology enables us to automatically generate, through the Simulink™ platform, embedded code with its makefile realized models; to send the generated files via Ethernet to a remote device (On Board Computer, etc.) with operating system "Linux," Real-Time Linux, etc.; to compile the project through makefile directly on a remote system; and to run it, all in a fully automatic and user-friendly way.

N. Tancredi (✉)
MBDA, Rome, Italy
e-mail: nazario.tancredi@mbda.it

S. Alunni
CapGemini, Rome, Italy
e-mail: stefano.alunni@capgemini.com

P. Bruno
University of Rome La Sapienza, Rome, Italy
e-mail: bruno.piergiuseppe@gmail.com

© Springer International Publishing Switzerland 2016
P. Ciancarini et al. (eds.), *Proceedings of 4th International Conference in Software Engineering for Defence Applications*, Advances in Intelligent Systems and Computing 422, DOI 10.1007/978-3-319-27896-4_7

75

1 Introduction

The proposed system is a multiplatform system for rapid prototyping (*Rapid Prototyping*). This technology gives the opportunity to implement and immediately verify the algorithms, such as Guidance, Navigation, and Control, on cheap, small, and powerful boards. Using this methodology, a *Simulink™* modeler will be able to automatically generate code from the algorithm under development and to download it immediately in a remote platform connected to a board *BeagleBone©*, *BeagleBoard©*, *WandBoard©*, etc.

Once the algorithm has been downloaded on the board, it can be tickled through a "real-time" connection by another system. The chance to immediately verify the goodness of an algorithm is essential in the modern development process because it gives the opportunity to:

- Accept or reject the variants during the design phase;
- Find bugs or functional defects in the early phase of design;
- Evaluate the performance of generated code.

One of the main advantages of the proposed solution is the feasibility to **cross-compile** the model using any remote platform, completely separating algorithm designer and test engineer. The designer can develop and test algorithms in Windows platform, and he can also send the model code to the test engineer (into an equipped laboratory); the test engineer will evaluate the algorithm performance directly on a real system: the On Board Computer prototype (in a different OS like Linux, etc.), harmonizing the work organization.

The methodology is also included within a mature Model-Based Design process.

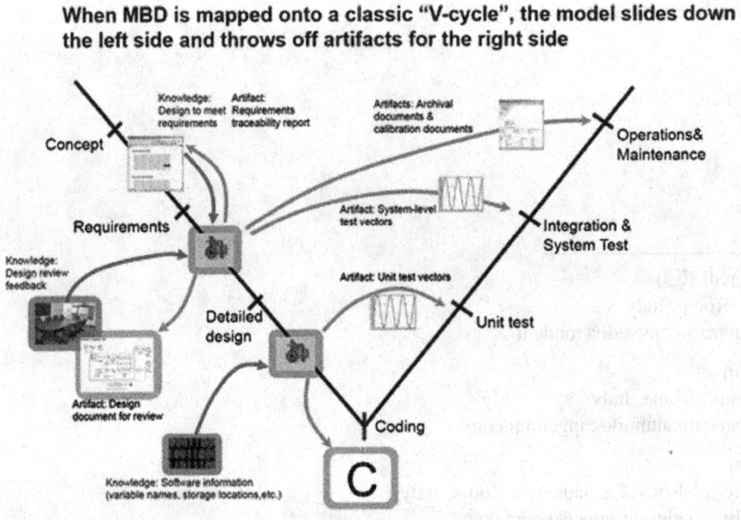

Fig. 1 MBD and V-cycle

The **Model-Based Design** yields to immediately test in reality what has been modeled, both during and at the end of the modeling process.

The described system plays an important role in this process mainly in the left side of the V-cycle development process (see Fig. 1).

2 Development Process

In Fig. 2 is shown the logic flow that characterizes the Rapid Prototyping process implemented. This approach allows to quickly test algorithm models, realized in *Simulink™*, directly on the real platform with *Real-Time Linux* operating system, being also able to compare the results with ones obtained from the simulation in a fully automatic and user-friendly mode.

Starting from the models realized in *Simulink™*, it is possible to generate automatically code in C/C++ language and the relevant *makefile* using the tool *Target Preferences* of the *Simulink™ Coder* library. This tool autogenerates code compatible for a specific target (Board, PC, etc.). The makefile created works only on Windows OS since *Simulink™* version used is installed on a PC with Windows operating system.

Fig. 2 Development process

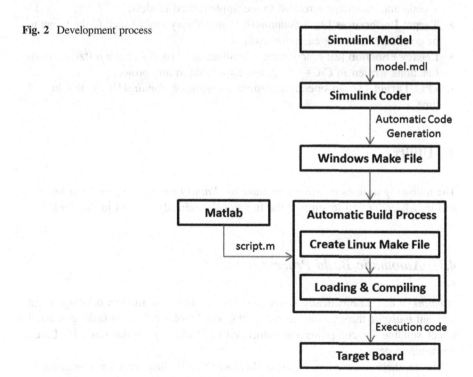

To solve this problem has been realized a *MATLAB™* script, "*Automatic Build Process*" in order to:

- create a new *makefile* compatible with the Linux operating system;
- create an archive *.tar* containing all generated files;
- submit the archive via *SSH* network protocol to a remote device;
- compile the project through previously defined *makefile* directly on the remote system (prototype board);
- run it on board.

3   MathWorks Toolset

In order to automatize algorithms coding and execution in a Remote Device with Linux Operating System (the Simulink model is defined in Windows) the following MathWorks tools have been defined:

- *Simulink™* [1]: software for modeling, simulation, and analysis of dynamic systems;
- *Simulink Coder™* [2]: *Simulink™* toolbox used to automatically generate C/C++ code and its *makefile* related to the implemented models;
- **Target Preferences** [2]: a *Simulink™ Coder* library tool to specify the target to be considered in autogeneration code;
- **Legacy Function** [2]: functionality for Simulink™ blocks creation starting from functions written in C/C++ language applicable to any project;
- **xPC Target** [3]: an operating system for running *Simulink™* models in real time.

4   Utility

The following utilities contribute to make the *Rapid Prototyping* process automatic and user-friendly and to support the functionality already present in *Simulink™*.

4.1   Automatic Build Process

Simulink™ can automatically generate C/C++ code of the models developed and relevant *makefile* through the "*Simulink™* Coder" toolbox. The *makefile* generated is not suitable for compiling the autocoded code directly on devices with Linux operating system.

Hence, there is a need to define a *MATLAB™* script that creates a new *makefile* compatible with Linux platform. This script called *Automatic Build Process*

Fig. 3 Automatic build process

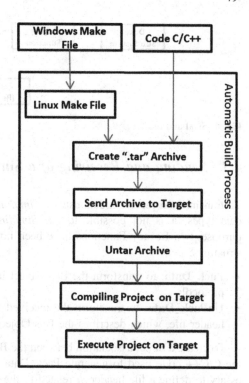

produces an archive *.tar* containing all files generated and the redefined *makefile*; download the file by *SSH* network protocol (via Ethernet) to a remote device; compile and run the project directly on it in a fully automatic and user-friendly mode. In Fig. 3 is illustrated the logic flow.

4.2 Use of Serial Port

Simulink™ can call functions written in C/C++ using Legacy Function in order to create new blocks suitable for any model. For example, it is possible to manage the messages sent and received from an external device directly using the drivers (written in C/C++) supplied by the manufacturer of the Equipment.

In specific case, it has been realized a *Simulink™* block that can manage, via a serial port RS-232, data received from an IMU system using the drivers provided by the manufacturer. In Fig. 4 is shown the flowchart related to the creation of a new serial port RS-232 *Simulink™* block.

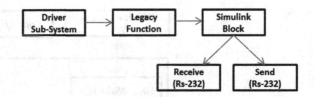

Fig. 4 Send and receive

4.3 Sending and Receiving of a Simulink™ Bus Object

To transmit and receive a data using a *Simulink™* Bus Object, containing different data types, it is not possible to use *Simulink™ Pack/Unpack* blocks. For this purpose, two Legacy Functions have been implemented to manage complex data structures:

- **Pack_Data**: to transform the Bus Object in a vector that can be sent via UDP protocol;
- **UnPack_Data**: to transform the received vector in a structure, defined in the header file, which describes the Bus Object.

To create a direct association between the Bus Object defined in *Simulink™* and the C/C++ code used to generate Pack_Data and Unpack_Data blocks, it is necessary to define a file header *.h* related to the signals to be sent or received.

To meet this requirement, a *MATLAB™* script has been defined with the purpose to automatically generate the files *.h* starting from file *.m* associated with the Bus Object. The file *.m* is created in *Simulink™* considering the Bus Object associated with the chosen model bus (virtual and non-virtual) to send or receive. Figure 5 shows the *Simulink™* blocks generated using two Legacy Functions.

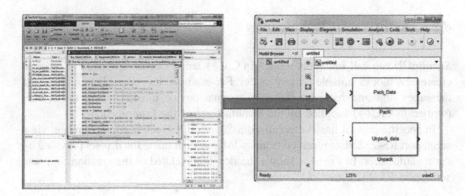

Fig. 5 Pack/Unpack

5 Applications

In the next sections are shown some examples related to this *Rapid Prototyping* process implemented using the tools and methodologies described above. For these applications have been used a *WandBoard* board and *FinxRTOS*.

5.1 *WandBoard*

The *WandBoard* is a development board equipped with "ARM Cortex-A9 Quad core" processor. This board has an internal memory of 2 GB DDR3 and has an SD card slot for loading the operating system directly on a microSD card [4]. Figure 6 shows the board's top side.

The WandBoard has been equipped with the EDM1-FAIRY expansion board (Fig. 7) for the integration of the WandBoard with different connection channels:

Fig. 6 WandBoard's top side

Fig. 7 EDM1-FAIRY

Fig. 8 Block diagram

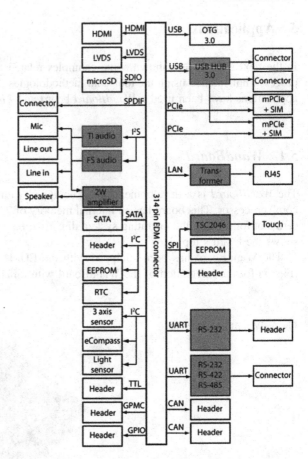

- serial ports RS-232/RS-422/RS-485;
- LAN, HDMI, USB 3.0, etc.;

Figure 8 shows the block diagram of the EDM expansion board. The operating system installed on the board is *Finx Real Time OS*.

5.2 Equipment Management: IMU

An application of the proposed process is the evaluation of the data provided by an IMU (inertial measurement unit—accelerometers and gyroscopes) connected directly to the WandBoard via a serial port, RS-232 [5]. Figure 9 shows the connection between the board and the IMU.

In Fig. 10 is shown the related *Simulink*™ model.

Fig. 9 WandBoard + IMU

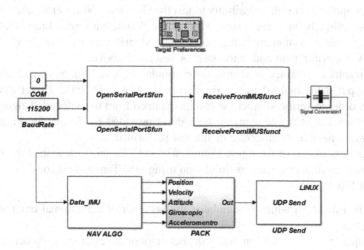

Fig. 10 *Simulink™* model for reading IMU data

In this case, the architecture defined and implemented in *Simulink™* is composed by three main elements (Fig. 10):

- Reception IMU data via serial port RS-232;
- Data processing using the Navigation algorithm (Kalman filter);
- Elaborate data transmission via UDP.

IMU data are managed by *Open_Serial* and *Receive_Data* blocks. These two blocks have been realized using Legacy Function considering the C code provided by the IMU manufacturer. The block *Open_Serial* opens the COM port of the board with a baud rate of 115,200. Moreover, it sends the file "descriptor" to the *Receive_Data* block (file that defines type and data structure that IMU is sending). The Navigation algorithm, with its Kalman filter, receives as input the data provided by the IMU (related to accelerometers and gyroscopes) and calculates the position, velocity, and attitude output data (the platform state). Then, the blocks *Pack* and *UDP Send* send the output data of the algorithm via UDP protocol from the board to the PC (*Simulink™* environment, Windows OS) where they will be displayed and saved. The output data so defined have been compared with the data obtained from the same model entirely evaluated in *Simulink™*.

The computational load occurred on the board is approximately the 5 % of its maximum one.

5.3 Guidance, Navigation, and Control Algorithms' Evaluation: Open-Loop Test

A further application is the possibility to test the Guidance, Navigation, and Control algorithms directly on a real system, such as the WandBoard, with Linux real-time operating system (computers in the loop). These algorithms have been developed in *Simulink™* environment and autocoded as described above.

The results so obtained in an open-loop simulation have been compared with the same output data obtained in the simulation (in *Simulink™*/xPC Target environment) in order to verify whether the results obtained from the autocoded algorithms executed on Target are comparable with those provided by simulation. Figure 11 shows the schematic architecture of the test performed.

In the bottom part of the figure (*Simulink™* xPC Target) is described the model of the entire system evaluated in simulation using xPC Target. This consists of three different blocks:

- **Environment**: in which is simulated the behavior of the external environment (physics models);
- **Equipment**: in which is simulated the behavior of the external equipment (IMU, GPS, etc.);
- **GNC**: it contains Guide, Navigation, and Control algorithms.

In a standard simulation loop, data provided by the sensors are the input for the GNC block which gives as output the deflections commanded to the control surfaces.

Fig. 11 Architecture

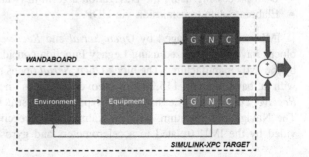

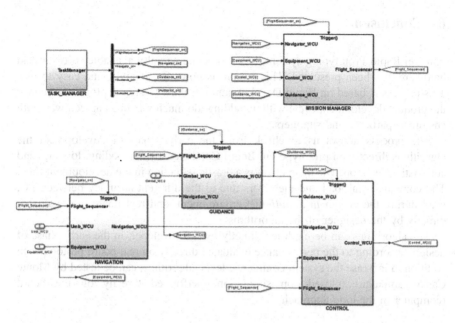

Fig. 12 Model GNC

In the test, the output data of the Equipment block in addition to being sent to the GNC block inside the simulation were transmitted in real time (xPC Target) via UDP to the target system (WandBoard) on which were loaded algorithms auto-coded, as shown in the upper part of the figure. At this point, the output data of the GNC block loaded on target were compared with those obtained by simulation. Figure 12 shows the *Simulink™* architecture of the GNC block.

As shown in the figure other than Guidance, Navigation, and Control blocks, there are two logic blocks, implemented in *Stateflow™* (state machine):

- **Task Manager**: it handles the timing of Guidance, Navigation, and Control blocks and defines the overall system scheduler providing signals Trigger to the GNC algorithms [6].
- **Mission Manager**: it handles the different mission phases. It has the purpose to generate events as a function of the fact that have occurred or not certain conditions.

Comparing the outputs provided by GNC on WandBoard (open-loop) with those provided by GNC simulated by xPC Target (closed-loop), the result was that the related outputs are identical, except for the different accuracy between PC and board.

6 Conclusion

A novel Rapid Prototyping system, based on commercially available software and hardware, has been presented. This system is powerful, flexible, and easy to use. This process is based on Rapid Development e Robust Design approach: to reduce the product development cycle without adding too much risk in an agreed way with customer, partners, and suppliers.

The process allows us to check from the early stages of development the algorithms directly on prototype On Board Computer, by autoçoding, loading, and automatic execution. The main effect is a drastic reduction in the development time: The computational load and the errors due to the different accuracy between PC, used during the design in *Simulink*™ environment, and real board are evaluated directly by the designer of the algorithms.

The algorithms can be modified directly by the developer, in the early phase of design, according to the performance evaluated directly on real board prototype. In addition to increase the system confidence level, algorithms can be tested by Monte Carlo campaigns (closed-loop simulation) performed directly on the board (computer-in-the-loop approach).

References

1. The MathWorks, Inc, "*Simulink® User's Guide*", 1990–2015 by The MathWorks, Inc
2. The MathWorks, Inc, "*Simulink® Coder toolbox User's Guide*", 1990–2015 by The MathWorks, Inc
3. The MathWorks, Inc, "*xPC-Target® User's Guide*", 1990–2015 by The MathWorks, Inc
4. Wandaboard, "*Wandaboard User Guide*", Wandaboard.org
5. Advanced Navigation, "*Spatial Reference Manual*", © 2014 Advanced Navigation Pty Ltd
6. The MathWorks, Inc, "*Stateflow® User's Guide*", 1990–2015 by The MathWorks, Inc

Pair Programming and Other Agile Techniques: An Overview and a Hands-on Experience

Marco Stella, Francesco Biscione and Marco Garzuoli

Abstract Although Agile programming methodologies are still relatively young, it is already possible to draw up an initial assessment of their potential outcome. This article provides a historical overview and a short description of these programming techniques and their relevant key tenets. Besides, based on several years of application, a practical point of view is provided, identifying the key points that demonstrated real long-term effectiveness as well as the unavoidable deviations from the theoretical model. At last, an interesting example of a practical application is given, regarding the development of a speech recognition-based squelch algorithm, suited for radio communications especially in the noisy HF range, useful for defense and security applications. In this case, not only has the application of Pair Programming led to the rapid integration of a newcomer in a well-established work group, but it also led to a simple and practical algorithm, which is reliable and easily implementable both in hardware and in software, with very promising performances: voiceXtract[TM].

1 Introduction

1.1 Prehistory

The first example of a programmable machine, intended as a machine which could perform a set of calculations defined by a user, can be historically attributed to

M. Stella (✉) · F. Biscione · M. Garzuoli
Larimart S.p.A, Via di Torrevecchia, 12, 00168 Rome, Italy
e-mail: marco.stella@larimart.it
URL: www.larimart.it

F. Biscione
e-mail: francesco.biscione@larimart.it

M. Garzuoli
e-mail: marco.garzuoli@larimart.it

© Springer International Publishing Switzerland 2016
P. Ciancarini et al. (eds.), *Proceedings of 4th International Conference in Software Engineering for Defence Applications*, Advances in Intelligent Systems and Computing 422, DOI 10.1007/978-3-319-27896-4_8

Charles Babbage [1, 2]. His idea was approved, but a first attempt at building the machine (the Differential Engine) was never completed due to the technological limitations of his times which led to unsustainable costs and resulted in funding cut. Between 1833 and 1842, Babbage tried to build a second machine (the Analytical Engine), which used input devices based on punched cards, an arithmetic processor, a control unit which determined whether the task was executed correctly, an exit mechanism, and a memory where numbers could be kept waiting for their turn to be processed. This device was the first computer in the world. Its concrete project came to light in 1837. However, due to difficulties similar to those encountered with the Differential Engine, it was never built. Lady Ada Lovelace, an English mathematician contemporary of Babbage, became very interested in Babbage's work. She actively promoted the Analytical Engine and wrote several programs for the Analytical Engine using what is now known as assembly language. However, those programs never run in practice on the machine. Ada Lovelace is therefore considered to be the founder of the science of programming, at least in its theoretical aspects.

After Babbage's machines, it took another century to build a new computer, in a more modern concept: WW2 urged scientists to create the first general-purpose computer in 1946: the Electronic Numerical Integrator and Computer (ENIAC), which was entirely based on electronic tubes.

Few years later, thanks to the introduction of the transistor, smaller, more powerful and much faster computers were made. Computer code grew accordingly in an attempt to exploit the full potential of the hardware they were made to run under and became more complex and diversified. Significant innovations in software were made, such as new programming languages. The software was developed according to the "code and fix" approach, which was based on writing software and then fixing it in sequential steps, until the maturity of the code itself was reached.

As long as the code was of modest size, this way of proceeding was manageable, but when systems became more complex, requiring the interaction of more people, and the needs of the market required shorter development times, the need of an effective method arose that would allow cooperation between different people and would reduce the "time to market" through a project management with well-established and well-organized objectives. In the late 1960s, it was noticed that in the management of complex software systems, it was not enough to be familiar only in programming and to be able to implement features in a certain programming language, and on the contrary, it was necessary to find a way to gain full control and management over all phases of the project, from its conception to its disposal.

The major cause of the software crisis is that the machines have become several orders of magnitude more powerful! To put it quite bluntly: as long as there were no machines, programming was no problem at all; when we had a few weak computers, programming became a mild problem, and now we have gigantic computers, programming has become an equally gigantic problem.

[Edsger Dijkstra, The Humble Programmer]

The concept of "software life cycle" was then introduced, which included all development phases: management, organization, and all the tools and methodologies for the creation and management of a software system.

1.2 The Waterfall Method

The first attempt to organize the software development activity was the introduction of the waterfall method, a model in which all the activities that are part of software life cycle are sequential, as shown in Fig. 1.

The development cycle can be divided into five basic stages, each using the result of the previous as its input. These phases are as follows:

- **Requirements**: All system requirements to be developed are detailed and defined in a formal and comprehensive way, leaving no ambiguity in the reader;
- **Design**: In this phase, system architecture, how functionality will be organized, and programming languages, etc. are chosen;
- **Implementation**: In this phase, the development of the software is carried out, which implements all the features described by the requirements, using the chosen programming language;
- **Verification**: During verification, software is checked for correctness and compliance with the requirements;
- **Maintenance**: Bugfix of the code after its field release.

This method prevents the parallelization of the phases, as well as keeping bugs under control at coding time.

As a result, this model leads to good results when the requirements of the system are clear and do not change during the development phase, since the change management cost grows exponentially during the development phase.

In addition, any problem blocking any of the phases causes the entire development process to stop.

Fig. 1 Waterfall model

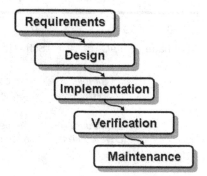

Due to the limitations of the traditional waterfall development method, the need for a different approach to the management of software development arose. A first attempt was conducted in the early 1960s, when, to reduce the development time of NASA Mercury Project, test procedures and software development were defined at the same time. There was no further evolution until a few years later, when the market needs justified its standardization. In particular, in the 1990s, the first principles of alternative development techniques were theorized, based on the iteration and parallelization of the different phases: The phases of development are independent from the others, less suffering the cost of changes.

1.3   The Rise of Agile Methodologies

To meet the new programming language requirements and the reduced life cycles of products, eXtreme Programming (hereafter XP) was created by Kent Beck in the 1990s.

Kent Beck, while working at the Chrysler management system payroll, in 1996 became project leader and began to define all the principles of XP. In 1999, he published "Extreme programming explained: embrace change" [3]. Thereafter [4], in 2001 Kent Beck himself, along with others, founded the principles of agile methodologies, and the Agile Manifesto [5] was released.

Figure 2 shows a time line that aims to highlight the key points that have contributed to the evolution of development techniques, as described above.

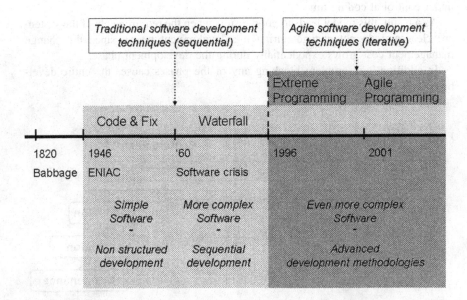

Fig. 2 Time line

Fig. 3 Agile methodologies'
life cycle

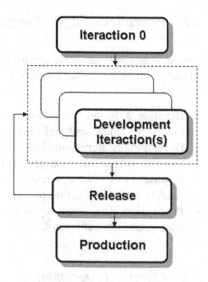

2 Agile Methodologies

2.1 Agile Manifesto

The principles of the Agile Manifesto [4–7] are the following:

1. Our highest priority is to satisfy the customer through early and continuous delivery of valuable software.
2. Welcome changing requirements, even late in the development. Agile processes harness change for the customer's competitive advantage.
3. Deliver working software frequently, from a couple of weeks to a couple of months, with a preference to the shorter timescale.
4. Business people and developers must work together daily throughout the project.
5. Build projects around motivated individuals. Give them the environment and support they need, and trust them to get the job done.
6. The most efficient and effective method of conveying information to and within a development team is face-to-face conversation.
7. Working software is the primary measure of progress.
8. Agile processes promote sustainable development. The sponsors, developers, and users should be able to maintain a constant pace indefinitely.
9. Continuous attention to technical excellence and good design enhances agility.
10. Simplicity (the art of maximizing the amount of work not done) is essential.
11. The best architectures, requirements, and designs emerge from self-organizing teams.
12. At regular intervals, the team reflects on how to become more effective, and then tunes and adjusts its behavior accordingly.

2.2 Principles of Agile Programming

Although they have different approaches to the software development management, all agile methodologies are based on the same principles and phases, as described in (Fig. 3):

- **Iteration 0**: In this first phase, setting up the team, acquiring the requirements, finding the funds, and performing the environment setup are performed;
- **Development iteration**: This phase requires several iterative substeps and foresees software development and testing;
- **Release**: In this phase, a release version of the software is issued, suitable for sending to the production phase;
- **Production**: This phase regards everything that occurs after the final software release, until its retirement.

2.3 eXtreme Programming

XP [3] is a particular incremental and iterative agile development methodology, in which the project is divided into subprojects. Figure 4 shows a basic diagram of XP phases and how they are linked to each other.

One of the major advantages of XP over traditional development methodologies is the lower impact of requirements change requests. Figure 5 shows, qualitatively, the impact of the changes among the development phases in the traditional cases as opposed to XP methodology.

The key tenets of XP are as follows:

- XP is based on the Pair Programming technique, which consists of programming in pairs, as described in the relevant section;
- Software development takes place according to the test-driven development (TDD) model. It is based on writing new functions only after the related test procedures definition. Moreover, test activities are performed involving both customer and programmers during each substep;
- In order to simplify the code, only functions that are actually needed are developed;

Fig. 4 XP main phases

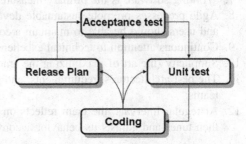

Fig. 5 Cost of changes

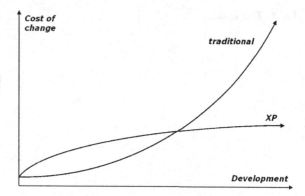

- Developers have to be prepared to make changes to the code at any moment; therefore, code must always be kept as clear and as simple as possible;
- In order to benefit from feedback during the requirements definition as well as during each release, meetings between the customers and the programmers are frequently held;
- Frequent releases of working and useful code are made at the end of each iteration.

The method was created with the aim to unify all concepts created in those years, which very often were based on common sense. XP is intended as a lightweight, efficient, low-risk, flexible, predictable, and scientific method.

2.4 Test-Driven Development

A programmer taking a TDD approach refuses to write a new function until there is first a test that fails because that function isn't present.

[Scott W. Ambler]

TDD is based on the test first design (TFD) concept that consists in coding the test functions before the actual functionalities. Figure 6 shows the block diagram of the TFD model.

TDD methodology adds to TFD the concept of *refactoring*, which traces the guidelines of the new features development. In particular, the development phases of the methodology are the following:

Add test: According to the specifications and requirements of the particular functions, the developer defines all the use cases that cover the related requirements and implements the code for the testing phase.

The execution of the test allows to:

- validate the test procedures themselves;
- verify that the new tests do not pass without the development of new features, thus verifying the actual need to develop new functionality;
- verify that the new features are not already available.

Fig. 6 TFD model

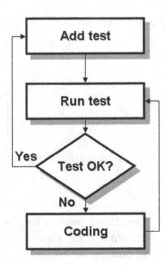

Write code: At the beginning of the development of a new feature, the related test will fail. The developer then proceeds to write new code just with the aim of passing that specific test. The code produced at this stage could possibly be very "rough" and unoptimized, but it will be refined in the next phases.

Run tests: When all the tests pass, the developer has such a level of confidence to consider the requirements met. Otherwise, the input code has to be further improved and possibly more code has to be written.

Refactoring: It is the main difference between TFD and TDD methodologies: In this phase, all the code added for the development of new features is reviewed with the aims of being renovated without changing its external behavior. In particular, one or more of the following operations are performed:

- Code sections that were added just to pass tests are moved to the appropriate position;
- Duplicated code sections in test code and in production code are removed;
- Objects, classes, module variables, and method names are reviewed for compliance with standards.

3 Pair Programming

3.1 Theory

The Pair Programming technique [8] consists of two programmers working together at the same time and on the same workstation. Each of them can take the role of:

- Actuator, who actually writes the code;
- Observer, who continuously checks the code written by the actuator, making contributions through strategic considerations, possible future opportunities, and possible problems.

The programmers must exchange their roles frequently, so that each of them can play each role seamlessly. Code writing must never depend on one specific programmer.

One of the main advantages of Pair Programming consists of continuous code testing by the observer. The code itself can then benefit from continuous comparisons and exchanges of opinions between the two programmers, ranging from the simplest to the most complex issues. The produced code, then, is much less affected by problems and much clearer than in the case of a single programmer. In addition, the Pair Programming contributes to the spread of knowledge of programming techniques: Topics that are known only to one of the two programmers are automatically disclosed to the other. Finally, the continuous and frequent alternation of roles prevents each of the two programmers to play a dominant role. There is a natural tendency to consider the role of the actuator as active and the role of the observer as passive, but they are not, or should not be, at all, and even the observer is called to actively participate in the development.

3.2 Application

Like all techniques, the Pair Programming has its shortcomings, mainly due to the context and to the affinity of the two programmers. In particular, a study by Cockburn and Williams [8] has quantified an increase of 15 % of the actual cost in terms of time to develop, thus discouraging its adoption by managers who do not see the advantages of a better final product. The study shows the economic, qualitative, and temporal analysis results of the technique, as compared to a single programmer. In Fig. 7, a comparison of the development time of different programs that are made over time by the same pair of programmers is shown.

The figure shows that within an initial period necessary to build up team spirit between the two programmers, usually very short, the development time of subsequent programs is almost identical (only higher than 15 %) to the case of development in a single-programmer mode. In economic terms, however, the development time corresponds to doubling the cost, because the total development time involves two people for the same program.

Moreover, as shown in Fig. 8, it is evident that in the same time of development, a pair of programmers write fewer lines of code.

The increase in the cost of development can be justified when you consider the resulting advantage on the final product: This method allows us to create programs with less errors, much of them caught by the observer at coding time, and therefore,

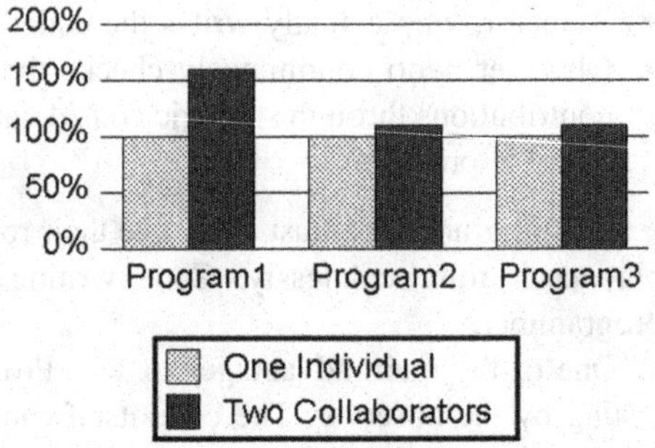

Fig. 7 Relative time: one individual versus two collaborators

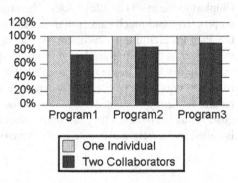

Fig. 8 Lines of code: one individual versus two collaborators

the investment cost of a higher development time will be compensated by a reduction in the maintenance cost [9–14].

4 XP and Pair Programming in Daily Routine

> You can get significant gains from parts of XP. It's just that I believe there is much more to be gained when you put all the pieces in place.
>
> [Kent Beck]

There are some limitations to the applicability and the success of agile development methodologies in general and of XP in particular: These limits may be procedural, technological, or simply related to the customer who prefers the release of the product in the specified date or is not equipped to participate at frequent meetings with the development group.

In particular, the company procedures can slow down to frequent releases due to a portion of unavoidable bureaucracy that steals valuable time to development.

Even the technologies used in the development may be an obstacle, such as the case of long-lasting compile processes of particular software, which makes frequent releases ineffective.

In any case, XP allows the adoption of even just a subset of its principles, limited to those aspects that actual cases can benefit from agile development.

After over ten years of application of agile programming techniques in the company, we can draw some conclusions from the daily way in which it was possible to manage the inevitable discrepancies between theory and practice. It must be noted that in the beginning, we started to apply these methodologies without being aware of Beck and his theories, the existence of which we realized only at a later stage. However, since the early days, there were many points of contact with Beck's theories, and the confluence of the agile development methodologies was a relatively natural consequence.

No doubt a strong push toward change and the dismissing of the waterfall model is the difficulty, at least in our area, to gather clear, complete, and sustainable requirements.

When the customer's requirements appear to be evidently unmatchable or unclear, we then identify some key points of the system around which to build an initial target Y_0, of which it is somehow possible to estimate the distance from the actual target X_0 necessary to meet the customer's satisfaction.

The largest initial effort is the implementation of the first objective, which has a decisive role in the proceeding of the activities. Touching with hands something close to the requirements generally has a clarifying effect on the user and on the developer, and contributes significantly to highlight and reduce the critical path. From this moment, the constant dialogue between end user and developer can be measured by estimating the distance between X_0 and Y_0. There is certainly a downside: The user may become aware that some requirements are not as important as planned and others may be more appealing. This is facilitated by an evolution from X_0 to X_1, and X_2 and then so on.

Very rarely a system is developed from scratch, it is instead much more frequent the case of adding functionalities or services to an existing base. While this situation may appear more favorable ("80 % of the system already exists, and you just need to develop the remaining 20 %") it must be said that, more often than not, the new services are generally incompatible with the rest of the system ("but that 20 % of system is going to disrupt the existing 80 %").

The authors have examined the case of the development of voiceXtract™, an algorithm to distinguish voice communications from noise in HF radio communications. The algorithm should have to be able to run in real time on an already consolidated system, without interfering with existing functionalities and operations. As usual, Pair Programming was adopted along with several principles of agile development methodology.

A team has been formed to prepare the requirements consisting of a senior designer and a junior designer. Another team, a supervisory team, consisting of two expert resources of radio communications was assigned the role of testing the system. Requirement definition has led to identify some basic criteria for the identification and isolation of the speech signal as opposed to noise. The requirements were translated to a block diagram which included signal analysis both in frequency and in time domain. In particular, a system to detect breathing pauses between one phrase and the other, as well as micropauses between a word and the other, was devised and defined. The supervisory team identified some algorithm test scenarios, independent of its implementation, which was assigned to the development team.

During the development, meetings were held almost daily for the definition of the path to follow, to update the outcomes, and to find and consolidate the ways to analyze the performances. In this case, the supervisory team played the role of the end user.

Note that the junior designer in the development team was entirely unskilled about both the existing system and the software development system to be used for coding the algorithm. Nevertheless, working near the senior designer helped him to catch up with the gaps very quickly and made possible to alternate the roles of observer and actuator between the two since the very early phases of the development. On the other hand, test management at encoding time proved to be more difficult, and the team therefore opted for a tighter control by the supervisory team to limit the occurrence of bugs during encoding.

Depending on the identified features, use cases were defined and automatic procedures developed for the analysis of sample waveforms and the outcome results.

In compliance with the 20–80 rules, the first 6 releases out of 27 totally (or 22 %) produced the most part of the overall effort to reach a preliminary system to be tested for final performances. In the first phase, a peak in developed code size is due to encoding in a sequential manner, without any optimization, just to pass the tests. This first step is essential to check the quality of the analysis and verification criteria, without worrying too much about the performances.

Immediately afterward, an optimization phase led to a significant reduction in the code size, thus obtaining the very same functions in a shorter time.

During the development, as the confidence on the final product increased, a major change in the project became necessary. Just as if it were a customer new requirement, it was then put in place.

Although the cycles of releases and iterations (Release Plan and Iteration Plan) were not rigidly defined, several releases were carried out. On three occasions, major software refactorings were undertaken: The first refactoring led to a rationalization of the code by discarding all the functions that demonstrated to be unnecessary, especially thanks to the greater confidence gained on the capabilities to implement (for instance, the DFT algorithm was replaced by a much simpler bank of IRR filters); in the second case, occurred at approximately one-third of the development, some data structures were trimmed down or even eliminated as they were no longer needed after the optimization. At this point, however, the two teams agreed that the performances of the algorithm were still too far from being satisfactory.

It was therefore necessary a radical change of strategy that forced the development team, compelled by the unsatisfactory results of the supervisory team, to redesign the time-domain analysis block. This change was hardly compatible with the data structures already optimized for the previous method. Rather than forcing the system to accommodate the new method, the development team decided to remove all variants hitherto entered and to reassemble them differently. In the vicinity of release 18, see Fig. 9, the code has been rewritten to obtain a fresh refactored environment to accommodate the new data structures, and signal analysis blocks were removed. This phase, as evidenced by a valley in the diagram below, was necessary to allow for non-regression tests of the algorithm host system.

Figure 9 shows the trend of the development of the code, and in particular of its size, during the development of voiceXtract[TM].

The new approach proved to be successful since the first full version, on release 20. At last, the performance of the system was satisfactory, both in its ability to identify the speech signal and in terms of performances. Figure 10 shows the trend

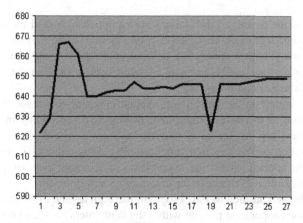

Fig. 9 Source code size versus voiceXtract[TM] releases

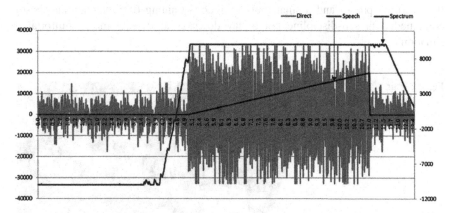

Fig. 10 Algorithm evaluation under heavy noise

of some of the parameters of the algorithm (spectral confidence and time-domain confidence) applied to a recording of a radio communication with S/N ratio of less than 5 dB.

The following development releases, made much easier by the new data structures, led to simpler and clearer code for the same performance. Despite the "performance crisis" encountered along the project's development, and the resulting change of approach, the time delivery constraints were met and it was not necessary to increase the pace of the work of testers and developers beyond limit (all people kept their sustainable pace). Furthermore, at the end of the project, the junior designer was able to operate in autonomy on the system and was successfully integrated in the group and the same algorithm has become the subject of a patent application.

Figure 11 shows two working stations used for development activities in our department.

As shown in the figure, all the stations are equipped with dual monitor and the desk is wide enough to easily accommodate two people sitting side by side that can comfortably look at both monitors and have access to the keyboard and the mouse: Both programmers can easily alternate between coding (just passing each other keyboard and mouse) and observing.

Pair Programming on the project was applied in senior–junior mode, especially in the first phase of the development, that mainly involved coding by those who thoroughly knew all the technical aspects and development strategies for the optimal realization of the required functions. Pair Programming in junior–junior mode was also applied, under the supervision of a senior member until all the juniors reached sufficient maturity for autonomous development.

The consolidated attitude to Pair Program, the high degree of harmony, and the spirit of extreme cooperation present within the department have also contributed to the success of this activity. But, of course, it is not always that easy.

It is clear that, as in all things, working this way requires both an initial effort (the learning phase) and a small but continuous training (the maintaining phase) combined with self-discipline not to lay down to apparently more comfortable situations.

Fig. 11 Pair Programming-friendly workstations

5 Conclusions

If there is no price to be paid, it is also not of value.
[A. Einstein]

Is it necessary to fully implement to the letter all tenets of agile methodologies to achieve a real business benefit? Paraphrasing an example of Kent Beck himself, the benefits of agile methodologies are like running on a bike. There is undoubtedly an initial price to pay to learn how to stay balanced. If this stage is not reached, it is not even possible to start talking about agile methods. But once you gain the balance, muscle memory is acquired that is never forgotten. Once on the saddle, it is not necessary to become a racing champion, and it is easy to see that those who still keep going by foot are doomed to lag behind. So it is possible to obtain sensible benefits from agile methodologies even without having to apply the theory in full.

In particular, Pair Programming and the application, even partial, of TDD criteria (applying specific tests only during the verification phase) can bring huge benefits as compared to programming alone, both in terms of the quality of the end product and in terms of a higher team spirit.

References

1. http://www.charlesbabbage.net/#. Accessed on 15 April 2015
2. http://www.computerhistory.org/babbage/adalovelace/. Accessed on 15 April 2015
3. Kent B (1999) Extreme programming explained: embrace change. Addison-Wesley, USA, 5 Oct 1999
4. http://www.agilealliance.org/. Accessed on 15 April 2015
5. http://agilealliance.org/the-alliance/the-agile-manifesto/the-twelve-principles-of-agile-software/. Accessed on 15 April 2015
6. http://www2.mokabyte.it/cms/article.run?permalink=mb165_metodoagile-1. Accessed on 15 April 2015
7. Royce WW (1970) Managing the development of large software systems. In: Proceedings of IEEE WESTCON, IEEE Press, (August 1970) Reprinted in Proceedings of international conference software engineering (ICSE) 1989, ACM Press, pp 328–338
8. Cockburn A, Williams L (2001) "The costs and benefits of pair programming", extreme programming examined. Addison-Wesley Longman Publishing Co., Inc., Boston, MA, USA, pp 223–243
9. Williams L, Kessler RR (2000) The collaborative software process. Doctoral Dissertation, The University of Utah
10. http://agiledata.org/essays/tdd.html. Accessed on 15 April 2015
11. Schümmer T, Lukosch S (2009) Understanding tools and practices for distributed pair programming. J Univers Comput Sci 15(16):3101–3125
12. Lui KM, Chan KCC (2006) Pair programming productivity: novice–novice vs. expert–expert. Department of Computing, The Hong Kong Polytechnic University, Hung Hom, Hong Kong
13. Dijkstra EW (1972) EWD340: the humble programmer. ACM Turing Lecture
14. Singh R (2008) International standard Iso/Iec 12207 software life cycle processes. Federal Aviation Administration, Washington, DC

Expressing, Managing, and Validating User Stories: Experiences from the Market

Carlo Pecchia, Marco Trincardi and Pietro Di Bello

Abstract Agile methodologies for software development favor customer involvement and thus a rapid feedback cycle on realized product increments. Such involvement is implemented in the activities around requirements (elicitation, analysis, development, management, change, validation), which in turn are often sustained by—and expressed in—"user story" format. This paper aims to show our experience in developing software system representing functional requirements mainly with "user stories," and capturing also nonfunctional requirements (e.g., availability, security) in demanding domains. This paper starts defining what a user story is, how we write and test it, and what are main differences compared to "traditional" documented requirements and use cases. Then, it focuses on techniques we use for splitting and grooming, and how we transform a linear backlog into a multidimensional Story Map that help us to manage size and complexity.

1 Introduction

In this paper, we want to shortly introduce our development process, then focusing on User Story (what they are and how we use them), and finally we describe main benefits we observed from our experience.

C. Pecchia (✉) · M. Trincardi · P. Di Bello
XPeppers srl, Trento, Italy
e-mail: carlo.pecchia@xpeppers.com

M. Trincardi
e-mail: marco.trincardi@xpeppers.com

P. Di Bello
e-mail: pietro.dibello@xpeppers.com

© Springer International Publishing Switzerland 2016   103
P. Ciancarini et al. (eds.), *Proceedings of 4th International Conference in Software Engineering for Defence Applications*, Advances in Intelligent Systems and Computing 422, DOI 10.1007/978-3-319-27896-4_9

2 Background

In developing products (systems), we use the agile framework known as Scrum [1]: an iterative approach where increments of functionality are implemented at each iteration. Beside Scrum, we make full use of Extreme Programming practices [2]: pair programming, test driven development, continuous integration, frequent releases, and so on. Each iteration starts with a short planning phase where some "user needs"—in the form of "user story"—are picked, discussed, analyzed by developing team with the business representative (also known as product owner), and then included in the iteration for developing. At the end of each iteration, we show the added increments of functionality in a demo session, where the product owner can give feedback for approval and/or necessary refinements (note this is a suitable opportunity to refine and adjust end user requirements).

One of the key tools used in this process are *User Stories*.

3 Introducing "User Stories"

A *User Story* can be defined as a short description of what end user does in order to achieve his functions when interacting with the system being developed. Generally, they consist in few sentences and are expressed in the business language. Figure 1 shows some examples.

A common practice, on writing user stories, consists of declaring explicitly the *who*, the *what,* and the *why* [3], as shown in Fig. 2.

A "role" (the "who") in the system does something (the "what") in order to achieve some business benefits (the "why"). However, other formats exist [4] that one is widely accepted and used mainly for its succinctness and *completeness*:

Fig. 1 Format examples

s-001

As a *Pilot,*
I want to *edit Legs and Waypoints,*
so that I can plan accurately my missions.

s-101

As a *System Administrator,*
I want to *set all Unit Measures for the System,*
so that I can help pilots have a coherent set of data.

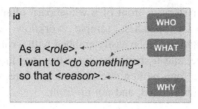

Fig. 2 General format

indeed such a format captures very essential aspects related to a functional requirement.

Some important aspects of user stories are listed below:

- the information carried by a single user story should be sufficient for its planning: can it be done in a single iteration or it should be split?
- a user story does not represent a formal requirement: it is a starting point for a real conversation with the end user (or his representative). Only with a real conversation, the development team can reach a (reasonable) shared understanding of what really are end user needs: a formal requirements document can be interpreted and understood in different ways by different people.
- being limited in scope, a single user story should be easily tested and verified.

3.1 The Process "Induced" by User Stories

To better understand the "process" enabled by user stories, let recall the 3C acronym [5]:

- **Card**—a conveniently short description of what functionality should be done from the end user standpoint.
- **Conversation**—a needed (face-to-face) conversation among development team about the user story.
- **Confirmation**—a common agreement on how to confirm that the user story is done.

Every single "C" is necessary in order to fully reach a shared understanding about what should be implemented and *how* it should be tested/verified. The agreement on the latter aspect (testing) is where generally important insights pop up; when we discuss on how to demonstrate the implementation of a single user story, we reason with (near) realistic data as examples. These discussions on "realistic examples" often permit us to find holes in story understanding, e.g., some external dependencies are not considered and data can present an unknown pattern.

Another important aspect, often misunderstood, is that "user stories" does not mean *lack of documentation*: is the *form* of the documentation that can vary.

When discussing a user story in front of a whiteboard often, we add some other information such as UI sketches, sample workflow diagrams. All those "low-fidelity" documentation can be easily grabbed with a camera and added to the user story as a reference, and it should be noted that it is far more expressive and cheap to produce than a long written documentation. Moreover, supporting documentation is written when needed, and not up-front.

3.2   Evaluating User Story

A widely accepted criterion applied in evaluating the goodness of a user story resides in the acronym INVEST [6]:

Independent
Each story should be completed (that means an increment of system functionalities can be done) independently from other stories; although some correlation, if not a strict dependency, may exist.

Negotiable
The scope of the user story can be adjusted when conversations—intended to clarify the user story—happen within development team.

Valuable
The story must deliver value to the intended business (directly or indirectly), if not one may consider to not developing it.

Estimable
The development team should be able to give an estimate, even a rough estimate, on the effort needed to implement the story. Otherwise, the business is not able to determine the effective cost/benefits ratio for the story.

Small
The story should be complemented in a single iteration; otherwise, it could be difficult to track progress or to be reasonably confident in having understood the whole story scope. As a rule of thumb, at least three stories should fit in an iteration.

Testable
When the user story is testable, the development team can confirm *when* it can be considered completed. Moreover, when those tests can be automated (that is a really desirable thing), they ensure that implementation of other user stories does not break those already done.

Applying a check on INVEST criteria, for each user story, gives to development team the confidence that the user stories "base" presents a sufficient level of quality that enables a smooth workflow. Of course, this "confidence" should not be guaranteed all up-front, but it can be achieved throughout iterations.

4 Testing User Stories

Together with each user story, we should describe its *acceptance criteria*: specific assertions that must be true for the user story to be considered "ready for the end user." It should be noted here an important aspect: These criteria are intended to provide enough guidance to the development team when implementing the story rather than detailing each specific test case. No one will prevent use to add—if needed—tests that verifies all conditions for the user story.

We also use TDD [7] to design and test user story implementation, but this is a different type of testing that works at very low level. Tests (first and foremost those covering the aforementioned acceptance criteria) are written before a very line of code is written for the user story. Once an automated test is written, we start coding for the story just enough to make the test pass, then we start to see improvements in the code (this is also known as "red-green-refactor cycle").

This approach gives us several benefits: (1) each story is covered by at least an automated test; (2) test automation enable us to run a regression test where other stories are implemented; and (3) we do not add functionalities not derived from acceptance criteria.

5 Differences from Requirements and from Use Cases

A general requirement is defined as [8]:

> A condition or capability needed by a user to solve a problem or achieve an objective.

Beside this definition, we found also derived requirements and nontechnical requirements.

User stories, being less structured, cannot be assimilated with requirements. That said, when there are mandatory reasons to document other types of requirements, we have to rely on extensive written documentation. In our practice, we tend to keep it as simple as possible in order to not waste time in low value added artifacts and activities.

Use Cases are an extensively used approach to expressing and documenting the functionalities of a system [9]. There is no direct correlation between user stories and use cases, because they work on a different level of detail. If we would like to find a similarity, we could see user story as part of use case scenario, but the latter is not needed to define the former.

In our practice, we tend to keep use cases (when present) for specific parts of the systems where its complex behaviors is clearly and fully expressed.

6 Managing Complexity

At the beginning, the scope of a story is so wide that it is not possible for the development team to implement the story in a single iteration and—more important—the end users voice cannot at the moment give more details, but only express an high-level need. In such a case, we talk about an *Epic* rather than a user story (Fig. 3).

Also, when we have a bunch of related stories, we can, for convenience, call the whole bunch as a *Theme*.

When a story is too big for an easy and concise understanding from the development team, we found convenient to split the story (see the example on Fig. 4). The general idea is, of course, to give the maximum value to the end user.

Some useful criteria and advices in splitting the stories are presented below:

- if the story describes a (complex or long) workflow, we should consider to split the workflow in several stories (e.g., first the start and ending points, then the intermediate states).
- if the story uses the same data type but with different interfaces, we can split the story by the "basic" handling of the data type, and then other stories for each interface.
- if the story is mainly related to nonfunctional specification (e.g., performance), we can split the story in a "simple" working version and then another one focusing on performance improvements.

Fig. 3 Epic samples

e-008

As a Pilot,
I want to be able to plan my mission with *aircraft*, so that ...

e-009

As a Pilot,
I want to be able to plan my mission with *UAV*, so that ...

e-010

As a Pilot,
I want to be able to plan my mission with *missiles*, so that ...

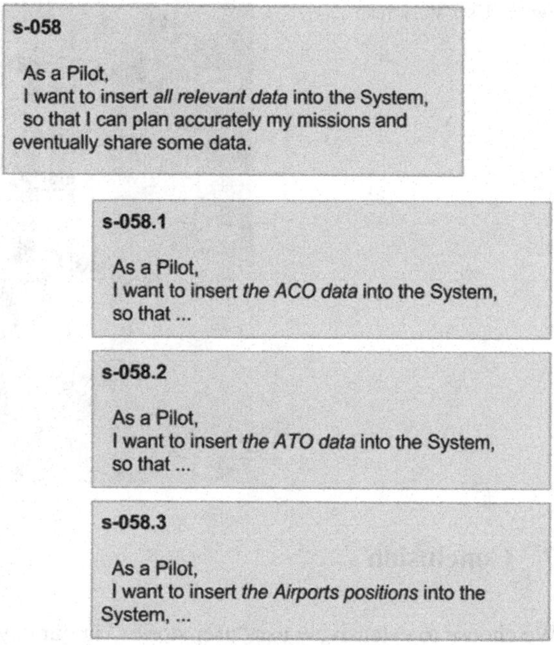

Fig. 4 Splitting sample

s-058

As a Pilot,
I want to insert *all relevant data* into the System,
so that I can plan accurately my missions and
eventually share some data.

s-058.1

As a Pilot,
I want to insert *the ACO data* into the System,
so that ...

s-058.2

As a Pilot,
I want to insert *the ATO data* into the System,
so that ...

s-058.3

As a Pilot,
I want to insert *the Airports positions* into the
System, ...

- if the story presents several variants of business rules, we can split the story in "main" and "secondary" rules.

It should be noted that the splitting activity is not all done up-front: it is an iterative activity. It is done at each iteration planning and in grooming sessions (see later on). We first split the story that gives the maximum value to the end user (the product owner voice here is essential) *and then* other big stories.

It should be noted that before starting a new iteration, we need a set of fine-grained user stories (allegedly those on the top of backlog); then, during the iteration, we spent time to clarify and reason more about other user stories. Indeed, we can deliver the maximum value to the end user in shortest possible time, so we start implementing a user story with a "just enough" understanding of it, investing other time on more examination when (and if) needed.

These "continuous" refinement activities on user stories are generally called "grooming": the set of user stories—generally called "the product backlog"—has to be continually maintained in order to be ready as possible for next iterations. To manage this set of stories at a different level of detail without loosing the big picture, we use a User Story Map [10]. This map is built incrementally and it represents a high-level view of activity done by the end user.

In our experience, the use of so-called low-fidelity tools such as post-it notes and whiteboards in managing user stories demonstrated a really effective way to (1) improve collaboration among people and (2) expose the whole set of features of a system/sub-system in a single, visible, place (Fig. 5).

Fig. 5 User story map

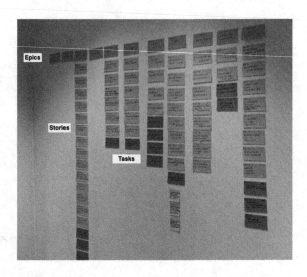

7 Conclusion

We choose to extensively use "user stories" in our development process for various reasons:

1. User stories avoid collecting too many details up-front, we delay such very moment when discussing "what" a user story really needs to be considered completed (and testable). This discussion arises when the development team considers including the given user story for the next iteration: until then we do not invest time and effort in gathering too many details, nor in estimating the *complexity size* of the story.
2. Although user stories represent, in our experience, an *effective* and *efficient* way to collect user needs, in some cases, we can—obviously—invest time in detailing some more elaborated "scenarios" when that helps avoid misinterpretations or helps clarifying system boundaries, constraints, and so on.
3. User stories carry the important "side effects" of the *need of a conversation* before considering its implementation. During this conversation(s), the development team can ask a wide range of questions to the business owner that represents the "voice of the end users." This is an important, and too often underestimated, aspect: face-to-face conversations with the help of low-fidelity prototype (post-it, whiteboard) are way too effective in clarity and expressiveness respect to—for example—a long and detailed requirements document (see also [11] for more insight about that).
4. Having had experience in sectors particularly concerned with quality, assurance, and urgency of delivery (mainly banks and financial), we believe that the approaches described here can also be applied for defense field. To this end, the considerations expressed in [12] can be considered fully supported.

References

1. Schwaber K (2004) Agile project management with scrum. Microsoft Press, Washington, DC
2. Beck K (1999) Extreme programming explained: embrace change. Addison-Wesley, Boston
3. Cohn M (2004) User stories applied: for agile software development. Addison-Wesley Professional, Boston
4. Five Ws technique on information gathering. http://en.wikipedia.org/wiki/Five_Ws
5. Jeffries, R (2001) Essential XP: card, conversation, confirmation. http://ronjeffries.com/xprog/articles/expcardconversationconfirmation
6. Wake B (2003) Invest in good stories, and smart tasks. http://xp123.com/xplor/xp0308/
7. Beck K (2003) Test-driven development by example. Addison Wesley-Vaseem, Boston
8. Chrissis MB, Konrad M, Shrum S (2011) CMMI for development. guidelines for process integration and product improvement. Addison-Wesley, Boston
9. Jacobson I (1992) Object oriented software engineering: a use case driven approach. Addison-Wesley, Boston
10. Patton J (2014) User story mapping. O'Reilly, Newton
11. Cockburn A (2000) Writing effective use cases. Addison-Wesley Professional, Boston
12. Lapham, MA (2010) Considerations for using agile in DoD acquisition. TECHNICAL NOTE CMU/SEI-2010-TN-002. SEI

Supplementing Agile Practices with Decision Support Methods for Military Software Development

Luigi Benedicenti

Abstract The literature shows that under certain conditions, traditional software development processes benefit from the adoption of structured decision support methods. Agile methods usually eschew this approach in favor of a collaborative decision-making structure. In the domain of defense software, however, a hierarchical structure is inherently present. Thus, the introduction of a more structured decision support method in agile development should lead to a higher level of comfort with the products built this way. This paper provides the foundation to adopting decision support methods as derived from past experiences at the University of Regina and grounds it in the defense software domain. The paper contains findings on insertion points for decision support methods and the methodology followed to evaluate which of these methods are the most appropriate. It also summarizes the lessons learned during the work conducted at the University of Regina and in selected industrial settings.

1 Introduction

Military software development covers a very wide spectrum of development environments, as it employs a combination of consultants and members of the military due to a number of nonfunctional requirements, frequently centered on security and confidentiality.

The two ends of the spectrum see a purely consultant-driven development, in which the military acts solely as customer; and a purely military-driven development, in which the military is in charge of all aspects of the development cycle and no external presence is allowed.

Military doctrine requires the presence of a chain of command. Agile software development is based on a somewhat different version of disciplined development

L. Benedicenti (✉)
University of Regina, Regina, Canada
e-mail: luigi.benedicenti@uregina.ca

© Springer International Publishing Switzerland 2016   113
P. Ciancarini et al. (eds.), *Proceedings of 4th International Conference in Software Engineering for Defence Applications*, Advances in Intelligent Systems and Computing 422, DOI 10.1007/978-3-319-27896-4_10

and naturally varies depending on the development process adopted for each development endeavor. This creates potential for a dichotomy of approaches, in which the inherent flexibility of agile software development may be seen to interfere with the chain of command.

This paper investigates this situation. The first step is a literature search to understand whether there is potential for such interference, and if so which aspect of such interference is more easily addressable. Then, a framework is proposed to implement a conflict resolution strategy based on decision theory. Finally, the framework is justified based on the results obtained in some industrial development environments in Canada. The main concepts and methods are then reported in the conclusions, along with the advantages and disadvantages of the proposed framework.

2 Background

Many approaches are possible to address the question of whether the chain of command approach may interfere with agile software development. In this paper, the chosen starting point is decision theory. This is a promising avenue for exploration because of its richness and its relevance to the essence of both military doctrine and software development.

Any creative activity requires decisions. Open-ended problems, such as software development, can be addressed and solved in many different ways, and any particular solution is the result of a large number of decisions. In most cases, these decisions are nontrivial: that is, it is not possible to accurately predict the outcome of the decision in advance.

Decision theory is a multidisciplinary field of knowledge concerned with the way in which decisions are made and alternatives are considered in the making of decisions. Decision theory encompasses multiple disciplines including Mathematics, Economics, Psychology, and Engineering [1].

Many researchers have conducted explorations of the need for decision theory in software engineering. Although a systematic literature review is beyond the scope of this paper, it is valuable to explore this area of the literature as follows.

According to a qualitative study of the impact of decisions on managers in software engineering production [2], hard decisions in software development carry strong emotional content and have the potential to destabilize the decision process in the long run. On this front, it is possible to argue that military-driven software development should have an edge on civilian-driven software development as the military have a set of focused techniques to deal with hard decisions and their consequences that are in general based on assiduous training and the chain of command.

The introduction of agile software development, however, complicates this situation somewhat. A recent study of obstacles to decision making in agile development identified six such obstacles. They are unwillingness to commit to decisions; conflicting priorities; unstable resource availability; and lack of: imple-mentation; ownership; and empowerment [3].

For agile projects including different stakeholders (as is usually the case for whole team projects), more issues arise, namely it is very hard to reconcile different points of view that often lead to different decision rationales [4]. All levels of decisions are affected: strategic, tactical, and operational. Companies involved in agile software development found out that Scrum development required the change from a rational decision model based on company hierarchy to a naturalistic model based on sharing [4]. As well, Scrum appears to offer a high level of empowerment to individual programmers versus more structured plan-based development pro-cesses [5]. In a military environment, this change may be seen as contrary to doctrine. Figure 1 provides a sample of issues that could be construed to this effect, and their generalization into a taxonomical description.

For the purposes of this paper, the NATO definition of military doctrine will be adopted: "Fundamental principles by which the military forces guide their actions in support of objectives. It is authoritative but requires judgement in application" [6]. This definition offers a point of reconciliation with decision theory, namely the requirement for decisions ("judgement") in the application of military principles.

In particular, NATO doctrine states, "Properly developed information require-ments ensure that subordinate and staff effort is focused, scarce resources are employed efficiently, and decisions can be made in a timely manner" [7].

Depending on the approach, decision theory can be classified as normative (prescriptive) and positive (descriptive) [1].

In this paper, the focus lies on prescriptive theory because of its more immediate practical applicability. Many prescriptive methods exist, such as the analytical

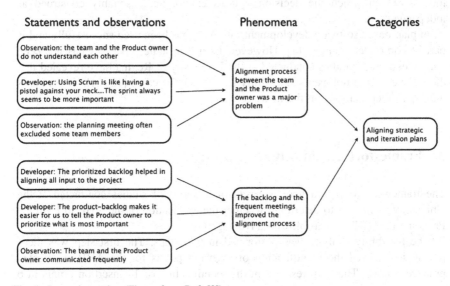

Fig. 1 Issue abstraction (Figure from Ref. [3])

hierarchy process (AHP), the analytical network process (ANP), the Stanford Normative School, value-focused thinking, and real options. Each method has more specific applications; in terms of reconciling decisions originating from multiple points of view or belief systems, the most promising approach at present seems to be either the AHP or the ANP.

The AHP is a systematic approach for problems that involve the consideration of multiple criteria in a hierarchical model. The AHP reflects human thinking by grouping the elements of a problem requiring complex and multiaspect decisions [8]. The approach was developed by Thomas Saaty as a means of finding an effective and powerful methodology that can deal with complex decision-making problems [9]. The AHP comprises the following steps: (1) Structure the hierarchy model for the problem by breaking it down into a hierarchy of interrelated decision elements. (2) Define the criteria or factors and construct a pairwise comparison matrix for them; each criterion on the same level is compared with other criteria in respect of their importance to the main goal. (3) Construct a pairwise comparison matrix for alternatives with respect to each objective in separate matrices. (4) Check the consistency of the judgment errors by calculating the consistency ratio. (5) Calculate the weighted average rating for each decision alternative and choose the one with the highest score. More details on the method, including a step-by-step example calculation, are found in [9].

The analytic network process is a more sophisticated version of the AHP [10]. In the ANP, the assumption of independence of options is relaxed. As a result, the hierarchical nature of the decision flow is no longer possible; hence, a network process must be adopted to reconcile the evaluations that need to converge to a decision. As the cost of using the ANP is higher than that of using the AHP, its adoption is more limited. In the proposed model, the ANP will be used only in those cases in which the decision options cannot be reasonably construed as independent.

In plan-based software development, the ANP has been used successfully and is part of corporate strategy [11]. However, its use in agile processes has not been widely explored. Studies conducted at the University of Regina have revealed that the AHP can be employed successfully in a variety of cases [12–14]. These cases will be further analyzed in the next section.

3   Framework Architecture

The framework presented here is based on a series of studies conducted at the University of Regina to determine whether the application of the AHP to Extreme Programming (XP) is feasible and advantageous.

The feasibility analysis was performed in two steps. The first step was theoretical: it involved the identification of insertion points for the AHP in selected practices of XP. This step resulted in the identification of 14 insertion points in 6 practices at various levels of abstraction: strategic, tactical, and operational. The results of these findings are shown in Table 1. Note that these are not necessarily all the insertion points that can be identified; these are simply the ones that pair most relevantly with AHP. Other insertion points can be identified in future work.

The abstraction levels in Table 1 deserve further explanation, as the definitions provided here come from business and differ from the military definition.

The **strategic** level involves the process. Therefore, strategic changes apply not only to one product, but also to an entire product line. This level of abstraction relates to the definition of strategy adopted by the military of "the utilization during both peace and war, of all of the nation's forces, through large-scale, long-range planning and development, to ensure security and victory" (*Random House Dictionary*).

The **tactical** level involves the project. Thus, tactical changes apply to a single product. Tactical changes can occur between product versions, for example, or if the situation changes (e.g., team composition). This definition is in general coherent with military theory, that defines tactics as (for example) "[t]he level of war at which battles and engagements are planned and executed to accomplish military objectives assigned to tactical units or task forces. Activities at this level focus on the ordered arrangement and maneuver of combat elements in relation to each other and to the enemy to achieve combat objectives" [15].

The **operational** level involves the iteration in a project. Thus, there is an expectation that changes will occur in every iteration. Strategy and tactics may change too, of course, but not in a steady state situation. The operational level still relates to tactics in a military theory context. Note that this concept is distinct from the term "operations" used for example in "joint military operations" as it is a much less general and strategic level of abstraction.

Table 1 Application of AHP to XP

Practice	Insertion point	Abstraction level
Planning game	Direct prioritization	Operational
	Requirements selection	Tactical
	Story estimation	Strategic
Simple design	Design tools	Strategic
	Cards selection	Operational
System metaphor	Metaphor selection	Tactical
Pair programming	Compatibility index	Operational
	Select rules	Tactical
Refactoring	Internal quality ranking	Tactical
	External quality ranking	Tactical
Test-driven development	Release decision	Operational
	Automated or manual test decision	Tactical
	Automated test candidates	Tactical
	Automation tool selection	Strategic

The second step for the feasibility analysis was to test each insertion point through a study. This was accomplished by embedding the study in a graduate course. The course enrolment was 12. Students were evenly spread between Master's and PhD levels. No student had been previously exposed to XP, and thus the learning curve for it was expected to be similar. The project lasted for 12 weeks of which two were for preparation, and the remaining 10 weeks were spent in five iterations of two weeks each. The customer was purposely chosen outside of the academic environment to provide better relevance to the project.

The study proved that it was possible to adopt AHP in all the insertion points, thus confirming the feasibility of the chosen approach; but there was still no appraisal of the benefits this would bring to the development process.

To determine whether there was a benefit introduced by AHP in XP, another study was conducted involving industry. Because of the difficulty involved in controlling the factors in an industrial setting, it was decided to adopt a qualitative approach to the study, which would sacrifice a more precise determination of the effect size for a more generalizable result.

The study involved three medium-sized companies located in Canada. Each company devoted six developers to the study and tested six of the 14 insertion points: compatibility index, select rules, internal quality ranking, external quality ranking, automated or manual test decision, and release decision. More information on the projects appears in Table 2. The names of the companies have been omitted to comply with their request for anonymity.

The results of the study confirmed the existence of a benefit in inserting AHP in these insertion points. The benefit was described in a qualitative way, and as such it was not possible to determine a quantitative indicator. Overall, the companies remarked that conflicts had been greatly reduced, and the quality of the software produced was in general higher than in their previous experiences. However, this benefit came at a cost: the time to make decisions with AHP increased geometrically with the number of criteria and alternatives considered. Companies also remarked that the integrity of their software development process was maintained.

Companies A and B were using Scrum, which allowed us to understand how the chosen insertion points would work in this case. For Company B, XP practices were already part of their development process, and Scrum was used to plan releases. In

Table 2 Industrial study parameters

Name	A	B	C
Application	Trouble ticket system	Order management system	HR gift card system
Programming language	JSF2/EJB3	Java	C#
Iteration duration	2 months	2 weeks	1 month
Development process	Agile/Scrum	XP/Scrum	XP
Programmers experience	14 years	10 years	12 years

this case, it was possible to use the insertion points with little adaptation, as the planning process in Scrum is relatively similar to the planning process in XP.

Company A did not use XP in their development process, so it was necessary to adapt the insertion points appropriately. However, Company A still used pair programming, which made it possible to use the pairing decisions in the same way as with Companies B and C. As pair programming is still lagging behind in terms of adoption (see, for example, [16]), the corresponding insertion points will not be adoptable by some.

The results obtained indicate that the proposed insertion points for AHP and the counting rules associated with them are a valid and beneficial framework for small-to-medium companies with some agile experience developing business systems.

The additional information provided above helps in tailoring the framework for agile development in military environments or for military software. Naturally, this information is only the foundation of a full investigation, but it provides some useful initial parameters.

Although these insertion points have worked well with AHP in the industrial studies, this is because in most cases, the reconciliation of options did not create any superposition of criteria, as all options were in fact independent. As well, it was possible to agree on an essentially unrelated set of criteria. Should these conditions not longer be met; it would be necessary to adopt the ANP, a more powerful version of the AHP, as previously indicated.

Thus the robustness of the framework would be increased by the addition of the ANP as an option for each insertion point. The ANP introduces the following advantages to the framework. First, it is more comprehensive than the AHP: it can deal with inconsistencies or lack of independence of options. Second, it does not introduce a predetermined hierarchy of options, potentially avoiding decision bias. Third, it is more comprehensive, and thus more suitable to decisions where human factors play a role.

The ANP has one major disadvantage, namely it is more computationally intensive than the AHP. Therefore, the use of ANP must be chosen only in those situations in which the AHP is not able to conclude. In the framework presented above, the ANP is most useful at the strategic level, whereas it is unfeasible to adopt it at the operational level. The tactical level should be employed on a case-by-case basis.

There is one additional component of the decision framework hereby introduced. Since the ANP and the AHP are an integral part of the framework, it follows that there will be at least one individual who will have received more in-depth training on these decision methods. This is, after all, part of the cost of ingraining the framework into the development process. Whether the individual in question is a manager, a project leader, or a Scrum master, this person has the possibility to decide and apply AHP or ANP to impromptu decision crises not covered by the framework.

For example, let us assume that a group has been developing an application for a few months using XP and that the customer, all of a sudden, must make a radical change to the requirements and thus set back development several weeks at least. In that case, an AHP framework could be set up to determine the criticality of the changes and the subsequent course of action. Similarly, make or buy decisions could be addressed with the AHP or, more likely, the ANP.

This way of reusing decision framework components is typical of agile methods and resonates well with independent, empowered groups. Paradoxically, however, it also works well in disciplined processes because it becomes a method for which specialized training can be offered, thus providing a systematic problem-solving avenue.

4 Conclusions

This paper contextualizes the application of the AHP and ANP to agile processes in the defense software domain in a general sense by introducing a decision-making framework which, when added to the development process, is able to bridge the requirements for empowerment that agile methods bring with the structural requirements of military doctrine and military theory.

The advantages of adopting the framework presented here are multiple:

- A standardized decision process is put in place, which helps structure the agile process while maintaining its key flexibility;
- Radically different points of view can be reconciled in a unifying process that avoids the sense of loss traditionally felt by manager (see [2]);
- The quality of development is positively affected;
- A systematic decision method can become part of standard manager/officer training;
- Team engagement grows higher because decision-making is perceived as collaborative and effective, enhancing the team's sense of empowerment.

These advantages, however, come with some trade-offs:

- The AHP and, even more so, the ANP are computationally intensive algorithms; thus, they should be employed only when necessary;
- Working protocols and training procedures might need to be tailored to make space for the AHP/ANP methods;
- The framework presented needs to be integrated with the chain of command, to ensure swift execution and acceptance.

It is also important to remember that the results obtained and presented here are not exhaustive. Ideally, future work would require a more thorough investigation of the quantitative aspect of the benefits introduced by the presented framework, especially in the defense software domain.

Acknowledgments The author wishes to acknowledge the contribution of the National Science and Engineering Research Council of Canada to this research.

References

1. Hansson SO (2015) Decision theory: a brief introduction. Royal Instititue of Technology, Sweden. Published online at http://people.kth.se/ ~ soh/decisiontheory.pdf. Retrieved 10 Mar 2015
2. Colomo-Palacios R, Casado-Lumbreras C, Soto-Acosta P, García-Crespo Á (2011) Decisions in software development projects management: an exploratory study. Behav Inf Technol
3. Drury M, Conboy K, Power K (2012) Obstacles to decision making in Agile software development teams. J Syst Softw: 1239–1254
4. Moe NB, Aurum A, Dybå T (2012) Challenges of shared decision-making: a multiple case study of agile software development. Inf Softw Technol 54:853–865
5. Tessem B (2011) An empirical study of decision making. Participation and empowerment in Norwegian software development organizations. Agile processes in software engineering and extreme programming. Lecture notes in business information processing, vol 77, pp 253–265
6. AAP-6(V) NATO glossary of terms and definitions
7. (AJP)-3 (B) Allied joint doctrine for the conduct of operations
8. Tiwari N (2006) Using the analytic hierarchy process (AHP) to identify performance scenarios for enterprise application. The computer measurement group
9. Saaty T (1980) The analytic hierarchy process. McGraw-Hill, New York
10. Saaty T (1996) Decision making with dependence and feedback: the analytic network process. RWS Publications, Pittsburgh
11. Hedstrom J (2010) Project portfolio management using IBM rational focal point. IBM Report. IBM
12. Alshehri S, Benedicenti L (2014) Ranking and rules for selecting two persons in pair programming. J Softw (JSW, ISSN: 1796-217X), 9(9)
13. Alshehri S, Benedicenti L (2014) Ranking the refactoring techniques based on the internal quality attributes. Int J Softw Eng Appl (IJSEA) 5(1):9–30
14. Alshehri S, Benedicenti L (2013) Ranking approach for the user story prioritization methods. J Commun Comput 10:1465–1474
15. The United States Department of Defense Dictionary of Military Terms
16. Ambler S (2015) Answering the "Where is the proof that agile methods work?" Retrieved from 10 Mar 2015, http://www.agilemodeling.com/essays/proof.htm#sthash.qpoOJHAd.dpuf

Benefits of Open Source Software in Defense Environments

Daniel Russo

Abstract Even though the use of Open Source Software (OSS) might seem paradoxical in Defense environments, this has been proven to be wrong. The use of OSS does not harm security; on the contrary, it enhances it. Even with some drawbacks, OSS is highly reliable and maintained by a huge software community, thus decreasing implementation costs and increasing reliability. Moreover, it allows military software engineers to move away from proprietary applications and single-vendor contracts. Furthermore, it decreases the cost of long-term development and lifecycle management, besides avoiding vendor's lock in. Nevertheless, deploying OSS deserves an appropriate organization of its life cycle and maintenance, which has a relevant impact on the project's budget that cannot be overseen. In this paper, we will describe some of the major trends in OSS in Defense environments. The community for OSS has a pivotal role, since it is the core development unit. With Agile and the newest DevOps methodologies, government officials could leverage OSS capabilities, decreasing the Design (or Technical) Debt. Software for Defense purposes could perform better, increase the number of the releases, enhance coordination through the different IT Departments (and the community), and increase release automation, decreasing the probability of errors.

1 Introduction

This paper claims five main issues about the benefits of the use of OSS in Defense environments: Cost, Innovation, Delivery, Security, and Agility.

Some of these issues appear pretty straightforward, such as the cost issue. In fact, the use of OSS decreases (or eliminates) contractor's monopoly regarding the exclusivity of the required capabilities. Furthermore, innovation is fostered, since Defense officials have more time to focus on their core business and not to code

D. Russo (✉)
Consorzio Interuniversitario Nazionale per l'Informatica (CINI), Rome, Italy
e-mail: daniel.russo@unibo.it

© Springer International Publishing Switzerland 2016   123
P. Ciancarini et al. (eds.), *Proceedings of 4th International Conference in Software Engineering for Defence Applications*, Advances in Intelligent Systems and Computing 422, DOI 10.1007/978-3-319-27896-4_11

(which is done by the community). Similarly, also the delivery is improved, because military software engineers can focus on changes and integration of existing software capabilities, instead of having to redevelop entire software systems. Time management is optimized to deliver new capabilities which relate more with the core business. It has not to seem strange that also the security is enhanced through a peer reviewed and reactive community. Furthermore, a broad access to the source code enables software security even after the release. Likewise, software development is agile, since there is no copyright infringement (since the OSS is licensed by the Defense agency or the Government) so it is easy for all departments to use clones (piece of codes) to set up their own systems, without any copyright infringement. It is like a public library, where any official can just pick what he needs to build his system.

DevOps is a growing software engineering methodology to assure a better integration of the Development, Quality Assurance, and Operations Department, in order to optimize the quality and the number of releases. This is, for instance, crucial in a dynamic environment where features have to change very fast (such as a battle environment). Furthermore, optimizing the releases decreases also the Design Debt, which is very typical of Scrum methodologies.

After an explanation of the (ii) background, this paper points out that the OSS community can lead to enhance several benefits for software development (iii) and software optimization through the newest methodologies (iv). In conclusion (v), we suggest further research in reference to the European defense software development system.

2 Background

The debate about the use of Open Source Software (OSS) in military programs is not new in literature [1]. Passively managed closed government programs, developed mainly with a waterfall methodology, are less flexible and less efficient than open source systems. Thus, the Department of Defense (DoD) is trying to shift through a more open Software Development Approach [2]. Nevertheless, there are two main problems that arise: IPRs and National Security. Usually, the government does not have the intellectual rights to make it more open, having, maybe, only purpose rights but not unlimited rights. Furthermore, the government wants to maintain a National Security advantage by classifying it, in order to not permit to others to see/use it.

The real point behind these issues is the real trade-off, i.e., are the benefits greater than the drawbacks? The answer to this question cannot be univocal; it definitely depends from the organization and settings of the software design. For example, we see an excellent respond of using open architectures for modeling and simulation (M&S) at NATO level [3]. In particular for developmental environments, where interoperability plays a major role (e.g., NATO), open architecture seems to be a good solution in order to create a fully respondent and effective environment (i.e., M&S). Clearly, we have to distinguish the level of confidentiality and sensibility of every environment, but especially for interoperability open architecture seems to have a good respond. We find in the literature many case studies where openly shared systems address main issue of interoperability [4]. Using Commercial-Off-The-Shelf products (COTS), in defense environment, has relevant cost savings effects through economies of scale; this is especially for non-critical systems, such Special Test Equipment (STE) for testing military avionics equipment. Therefore, the potential cost savings due to COTS usage are proportionately greater in STE than in the higher volume avionics systems that are tested [5]. A second major benefit of using COTS products is that test system development schedule cycle time is greatly reduced [5]. Even though the realization of flexible and robust systems without a supporting architecture is difficult, requires significant system rework, and precludes the concept of a product line, still the use of COTS seems to be worthy [6]. The use of Agile development system is probably of the most appropriate organization of software development life cycle and maintenance, which has a relevant impact on the project's budget. In particular in defense environment, "sustainability" is a major issue [7]. Successful software sustainment consists of more than modifying and updating source code [8]. It also depends on the experience of the sustainment organization, the skills of the sustainment team, the adaptability of the customer, and the operational domain of the team. Thus, software maintenance as well as operations should be considered part of software sustainment [9]. We know, actually that the majority of defense system lifecycle costs are incurred after the system has entered operations. Operations and sustainment costs can easily reach 60 to 80 % of a weapon system's total lifecycle costs, depending upon the type of system and the duration of its employment [10]. DevOps, blurring the lines between software development and operations teams, pushing continuous integration even earlier in a product/system life cycle, seems a promising potential option for use in IT systems and weapon and logistics support systems [11].

3 Benefits of OTD

As it is known, the Open Technology Development (OTD) at the DoD has become reality and it is used to develop military software. Software developers of the community (i.e., not governmental officials) and governmental developers develop and manage collaboratively software in a decentralized way [12]. Thus, OTD is grounded on open standards and interfaces, OSS, and designs. Furthermore, collaborative and distributed online tools and technological agility enhances OTD.

These practices are proven and in use in the commercial world. Likewise non-military environments, the DoD is pushing for open standards and interfaces that allow software to evolve in a complex development environment. Therefore, using, improving, and developing OSS might minimize redundancy in the development and maintenance process, fostering the agile development of software. Not surprisingly the DoD uses OSS also for critical applications, considered as structural part of military infrastructure, especially in four broad areas: (i) Security, (ii) Infrastructure Support, (iii) Software Development, and (iv) Research [2]. Collaborative and distributed online tools are now widely used for software development. The private sector also often strives to avoid being locked into a single vendor or technology and instead tries to keep its technological options open (i.e., using OSS). Removing such OSS tools (e.g., OpenBSD) would mean to harm crucial infrastructure components on which network (i) Security relies on. Furthermore, it would also limit DoD access to the use of powerful OSS analysis and detection applications that hostile groups could use to help stage cyberattacks, as the general expertise in it. Finally, the established ability of OSS applications to be updated rapidly in response to new types of cyberattack would be harmed. This is in part because DoD groups use the same analysis and network intrusion applications that hostile groups could use to stage cyberattacks. The uniquely OSS ability to change infrastructure source code rapidly in response to new modes of cyberattack has been proven to be very effective [2]. Therefore, OSS has been proven to be reliable for many DoD's application, also in critical and sensitive ones, such as cyberattacks. Interestingly, from an IPR perspective, the GPL (the most common used license in the DoD) turns out to be surprisingly well suited to use in security environments. This is because of the existing and well-defined abilities to protect and control release of confidential information. The established awareness largely removes the risk of premature release of GPL source code by developers but, at the same time, developers make an effective use of the autonomy of decision typical of the GPL license. The (ii) Infrastructure Support depends on OSS, since OSS applications rely on the ability of the DoD to support Web and Internet-based applications. (iii) Software Development relies on the OSS community to grab from the large pool of software developers, also with specific skills in different programming languages, directly outgrowths of the Internet. Finally, (iv) Research benefits from OSS's little support costs. The unique ability of OSS to support sharing of research results in the form of executable software is particularly valuable for the DoD.

Like in any commercial-OTD environment, it is the software community that has the proper access to source code and designs documents across the company interacts with the company itself. This creates a decentralized development environment, leveraging existing software assets. Not surprisingly, OTD methodologies that have been used from OSS to open standard architectures have their most successful implementations from the direct interaction with the end-user community. The only way to make an OTD successful is, thus, the merging interest and inputs of both developers and users.

Briefly, we will now highlight the most controversial issues of OSS in security and defense environments.

3.1 Security Issues

A recent qualitative analysis showed that among software security professionals, OSS is perceived as a powerful defense tool against attackers [13]. Interestingly, no major issues were emphasized in the close vs. open source comparison of the two software paradigms in the context of vulnerability and risk management. One relevant professional affirmed that it is "...impossible to say which one is more secured and has less vulnerability, look just at Borland [closed source software] example that had a backdoor password for many years. It was discovered only recently" [13]. Among professionals, there is the common belief that the time to reach and fix the vulnerability in open source security software is much faster than closed source software, therefore more efficient. Nevertheless, it is also worthy to say that in their opinion most of the closed source software is generally better tested and contains less bugs and vulnerabilities. What it is interesting to notice is that having full access to the source allows an independent assessment of the exposure of a system, like any peer review system. Also the risk associated with using the system makes patching bugs easier and more likely and drives software developers to spend more effort on the quality of their code, in order to be not blamed by the community in which they have freely chosen to engage, with an easy software quality assessment [14].

Even if we cannot affirm that OSS is more valuable than close systems, we can at least state that it has, for many applications, the same dignity.

3.2 Cost Issues

Even if OSS is free, this does not mean that is has no cost. We can for sure argue that since it is breaking vendor's monopoly, it lowers lock-in costs. More in detail we can say that there are some cases where the use of OSS is cost beneficial, like in stable slow-growth environments with a large number of software installations the low purchasing and maintenance cost of OSS can result in savings and thereby increased profitability [15]. More in general we can state that adopting OSS lowers the cost and increases the operating efficiencies. Studies like Spinellis et al. point out that cost optimization is the main reason why large organizations switch from close to open software [15]. Some academics also argue that since the market will provide companies with a closed system which is most suitable for their needs in terms of flexibility, technological sophistication, or ability to adapt software to their specific needs, the major benefit of OSS is the price [16].

3.3   Innovation Issues

Open Source is basically a huge library where everyone could pick what he needs. Pieces of code from different programs can be assembled without having to invent a new system from scratch or break IPRs. Developers can rapidly assemble and modify existing systems and components, focusing their time and effort writing the code that takes standing capabilities to a higher degree, or combines already-existing components into one integrated system. Thus, programmers need to focus on changes and integration of new and critical software capabilities. This form of software reuse is called in literature software cloning [17].

3.3.1   Cloning

There is a relevant debate in literature about the advantages and disadvantages of cloning. In Table 1, we figure out the most relevant issues.

4   Sustainable Software Development

Open source is not a *panacea*. There is and will always be the need for software engineers to code, test, deploy, and operate. Open source represents a valuable tool to, e.g., overcome single vendor's lock-in and improve network security, among others. In a survey of over 400 business executives conducted by the IBM Institute for Business Value (IBV), it came out that even if software development and delivery are felt as "critical" by software houses, only 25 % believe their teams leverage development and delivery effectively [18]. IBM called this, "execution gap", which is basically the difference between the need to develop and deliver

Table 1   Advantages and disadvantages of code cloning

Advantages	Disadvantages
Clones are useful if different customers share similar requirements [23]	High maintenance costs [24]
Some programming languages encourage the use of tem plates, which result in software cloning [23]	Propagation of bugs: if a clone contains an error, it will spread rapidly over other parts of the program [25]
The use of clones can respond, sometimes, to efficiency requirements in the development [26]	Cloning discourages the use of refactoring, leading to a bad design of the system [27]
Using clones reduces the time required to develop a program [28]	Using clones increases the size of the code, leading to a less efficient system [29]

software effectively and the ability to do so. IBM concludes that this gap is causing missed business opportunities for the vast majority of companies. This image is useful to understand the centrality of a sustainable software development. Organizations that use OSS cannot override such issues. Therefore, it is important to understand the main topics of software development in OSS integration.

Software development methodologies appear to be more complex and mixed than just straightforward techniques, as the 8th *Annual Survey on the State of Agile* suggests with 55 % of prevalence of Scrum in Agile [19]. Many implementations in execution appear to be hybrids of Agile methods, with some traditional methodologies, such as Waterfall. Such a hybrid is usually called Water-Scrum-Fall, known also as a flexible development approach that includes both waterfall and Agile development principles [20]. The point is that organizations, usually, utilize Scrum software development techniques but employ traditional waterfall methodologies for non-development issues (e.g., planning and budgeting).

What often happens is a poor software design, caused by different factors, like business pressure, lack of deep system understanding by both developers and business, lack of documentation or collaboration. The technical (or design) debt can also be described as the gap between the current state of a software system and an idealized state, in which the system is perfectly successful in his environment [21].

Sustainable software development methodologies can narrow this gap. This is possible through a tight relationship and cooperation between customers and producer but also between the development and the operation department of a company. Agile methodologies give some important answers to the technical debt issue as also DevOps. Implicitly, a DevOps approach applies agile and lean thinking principles to all stakeholders in an organization who develop and operate. It balances development and operation concepts with the aim of changing cultural mindsets and leveraging technology more efficiently [22].

5 Conclusions

In this paper, we pointed out some major issues of OSS integration within a Defense Environment. We also highlighted that any integration cannot disregard from sustainable software development.

Future research could take into consideration some case studies within European Armies of OSS integration. Even if we have some research about the DoD, studying OSS implementation outside the USA could interesting to confront if the USA represents a specialty or they are in line with other Defense Organizations.

Acknowledgments The author would like to thank Prof. Paolo Ciancarini for his helpful remarks and continuous inspiration.

References

1. Scott J, Wheeler DA, Lucas M, Herz JC (2011) Software is a renewable military resource. DACS 14(1):4–7
2. MITRE Corporation (2003) Use of free and open-source software (FOSS) in the U.S. Department of Defense
3. Hassaine F, Abdellaoui N, Yavas A, Hubbard P, Vallerand AL (2006) Effectiveness of JSAF as an open architecture, open source synthetic environment in defence experimentation. In: Transforming training and experimentation through Modelling and Simulation, vol 11, pp 11-1–11-6
4. Sawilla RE, Wiemer DJ (2011) Automated computer network defence technology demonstration project (ARMOUR TDP): concept of operations, architecture, and integration framework. In: IEEE international conference on technologies for homeland security (HST), pp 167–172
5. Pizzica S (2001) Open systems architecture solutions for military avionics testing. IEEE Aerosp Electron Syst Mag 16(8):4–9
6. Henry M, Vachula G, Prince GB, Pehowich J, Rittenbach T, Satake H, Hegedus J (2012) A comparison of open architecture standards for the development of complex military systems: GRA, FACE, SCA NeXT (4.0). In: Proceedings—IEEE military communications conference MILCOM, pp 1–9
7. Defense Acquisition University (2013) Sustainment Archived References: https://acc.dau.mil/CommunityBrowser.aspx?id=677043. Accessed on 10 April 2015
8. Lapham MA, Woody C (2006) Sustaining software-intensive systems (CMU/SEI-2006-TN-007). In: Software Engineering Institute, Carnegie Mellon University. http://resources.sei.cmu.edu/library/asset-view.cfm?AssetID=7865. Accessed on 10 April 2015
9. Defense Acquisition University (2011) Integrated Product Element Guidebook. DAU, November 2011: https://acc.dau.mil/CommunityBrowser.aspx?id=496319. Accessed on 10 Apr 2015
10. Taylor M, Murphy J (2012) Colt. OK, we bought this thing, but can we afford to operate and sustain it? Defense AT&L: Product Support Issue, Mar–Apr 2012, pp 17–21
11. Regan C, Lapham MA, Wrubel E, Beck S, Bandor M (2014) Agile methods in air force sustainment: status and outlook (CMU/SEI-2014-TN-009). Software Engineering Institute, Carnegie Mellon University, USA
12. Herz JC, Lucas M, Scott J (2006) Open technology development roadmap plan, Apr 2006. Office of the under secretary of defense for acquisition
13. Silic M (2013) Dual-use open source security software in organizations e dilemma: help or hinder? Comput Secur 39:386–395
14. Hoepman JH, Jacobs B (2007) Increased security through open source. Commun ACM 50(1):79–83
15. Spinellis D, Giannikas V (2012) Organizational adoption of open source software. J Syst Softw 85:666–682
16. Attewell P (1997) Technology diffusion and organizational learning: the case of business computing. Organ Sci 3:1–19
17. Rattan D, Bathia R, Singh M (2013) Software clone detection: a systematic review. Inf Softw Technol 55:1165–1199
18. IBM (2013) The software edge: how effective software development and delivery drives competitive advantage. IBM Institute for Business Value: http://www-935.ibm.com/services/us/gbs/thoughtleadership/softwareedge. Accesed on 18 Apr 2015
19. VersionOne (2014) 8th annual state of agile survey. VersionOne, Inc.: http://stateofagile.versionone.com/. Accessed on 15 Jan 2015
20. West D, Gilpin M, Grant T, Anderson A (2011) Water-Scrum-Fall is the reality of agile for most organizations today: manage the water-scrum and scrum-fall boundaries to increase agility, Forrester Research

21. Brown N, Cai Y, Guo Y, Kazman R, Kim M, Kruchten P, Lim E, Maccormack A, Nord R, Ozkaya I, Sangwan R, Seaman C, Sullivan K, Zazworka N (2010) Managing technical debt in software-reliant systems. FSE/SDP workshop on the future of software engineering research, FoSER, pp 47–51
22. Swartout P (2012) Continuous delivery and devops: a quick start guide. Packt Publishing, Birmingham
23. Kim M, Bergman L, Lau T, Notkin D (2004) An ethnographic study of copy and paste programming practices in OOPL. In: Proceedings of 3rd international ACM-IEEE symposium on empirical software engineering (ISESE'04), Redondo Beach, CA, USA, pp 83–92
24. Monden A, Nakae D, Kamiya T, Sato S, Matsumoto K (2002) Software quality analysis by code clones in industrial legacy software. In: Proceedings of 8th IEEE international symposium on software metrics (MET-RICS02), Ottawa, Canada, pp 87–94
25. Johnson JH (1994) Substring matching for clone detection and change tracking. In: Proceedings of the 10th international conference on software maintenance, Victoria, British Columbia, Canada, pp 120–126
26. Kapser CJ, Godfrey MW (2008) Cloning considered harmful considered harmful: patterns of cloning in software. Empirical Softw Eng 13(6), 645–692
27. Lavoie T, Eilers-Smith M, Merlo E (2010) Challenging cloning related problems with GPU-based algorithms. In: Proceedings of 4th international workshop on software clones, Cape Town, SA, pp 25–32
28. Kapser CJ, Godfrey MW (2006) Supporting the analysis of clones in software systems: a case study. J Softw Maintenance Evol Res Pract 18(2):61–82
29. Koschke R (2008) Frontiers of software clone management. In: Proceedings of Frontiers of Software Maintenance (FoSM08), Beijing, China, pp 119–128

Agile Software Development: A Modeling and Simulation Showcase in Military Logistics

Francesco Longo and Stefano Iazzolino

Abstract As well known, Agile Software Development encompasses a set of methods and methodologies for complex and nondeterministic software development projects involving self-organizing and cross-functional groups. Therefore, as in the Manifesto for Agile Software Development, being adaptive, responsive, and cooperative as well as reliable are the underpinning requirements of Agile Methodologies whose inherent nature suggests that profitable results can be achieved when agile practices are absorbed into Modeling and Simulation projects. As a matter of facts, a lot of common ground can be found if well established M&S principles are evaluated against agile practices or vice versa. Indeed, agile software development, likewise simulation, relies on human and technical factors management along the project life cycle. Needless to say that it is not a surprisingly achievement, given that simulation building blocks are pieces of software components integrated with one another. However, it is worth noticing that simulation can provide the groundwork to assess and validate the effects of agile practices and in turn can substantially strengthen its methodological foundations benefiting from advances and best practices in software engineering. To this end, a case study in military logistics will be presented in this paper showing that high-quality results can be achieved applying agile techniques for simulation model development.

Stefano Iazzolino: Italian Army General Staff.

F. Longo (✉)
University of Calabria, Cosenza, Italy
e-mail: francesco.longo@unical.it

S. Iazzolino
Rome, Italy
e-mail: stefano.iazzolino@esercito.difesa.it

© Springer International Publishing Switzerland 2016
P. Ciancarini et al. (eds.), *Proceedings of 4th International Conference in Software Engineering for Defence Applications*, Advances in Intelligent Systems and Computing 422, DOI 10.1007/978-3-319-27896-4_12

133

1 Introduction

Logistics plays a key role in the whole chain of defense. According to Prebilic [1], logistic capabilities affect the availability of armed forces that can be employed in combat operations and as a result can set considerable limits of strategy.

Therefore, military logistic management requires robust as well as reliable approaches and tools aimed at reducing uncertainty while ensuring resources (i.e., means, materials, and platforms) availability and reliability regardless of distance and environmental conditions on national territories and abroad. However, military logistics planning is far from simple. Indeed, it involves collaboration among many organizational and informational entities that are geographically distributed [2], and, in addition, multiple variables and factors must be taken into account during decision-making processes.

Therefore, the goal of military logistics was to ensure the possibility of accomplishing a mission successfully supplying proper forces preferably with cost optimization. In particular, dealing with the logistics sustainability of a mission requires that logistic support to operations is properly planned and managed in order to run an OPORD (operation order) where logistics support units are suitably defined and allocated and as a result military operations are optimally sustained.

Needless to say that these aspects and their harmonization within the entire logistic system require a great deal of analysis, review, and testing since the very beginning of the design and engineering phase. Within this context, Modeling and Simulation is an innovative, fast, and powerful approach able to support at various levels the logistic chain management while overcoming practical limitations of traditional analysis [3, 4].

In this perspective, as shown in Slats et al. [5], logistic chain modeling is an enabling factor to improve logistic systems' overall performances but substantial advantages, as highlighted by Yücesan and Fowler [6], can be achieved when simulation is applied:

- Temporal Power. Simulation allows analyzing the behavior of a system over a wide time horizon [7]. It would be otherwise difficult due to the long-term effects that may be hardly predictable with traditional real-time investigation [8].
- Interaction Analysis. Agent-based concepts can be easily applied to analyze the dynamic behavior of supply chains. So that interactions, relationships, and dependencies among the system components and/or variables can be suitably taken into account and assessed (i.e., relationship between transport and lead times, as the human factor affects the efficiency of the entire process).
- Risk and prediction analysis. Simulation models allow evaluating the impact of new decision-making policies, structural rearrangements, or totally new solutions without making any physical change to existing systems and practices [9]. Thus, financial and physical risks connected to questionable choices [10] (i.e., warehouse disposal, activation of a logistics node, and stock levels decrease) can be substantially reduced.

- Repeatability. Simulation allows investigating a system behavior under different operational scenarios, or vice versa analyzing and comparing how various systems perform under specific conditions (i.e., it may be possible to test the configuration of an existing supply chain in different mission scenarios or evaluate a new one against several potentials alternatives in order to detect the one that best fits specific requirements set out by the context analysis [11]).

Tracking and monitoring. Each system component, recreated within the simulation model, can be methodically controlled and investigated. It turns out very useful for supply chain management and control and especially for military logistics.

From a methodological point of view, simulation can be implemented according to various approaches; for instance, Fujimoto [12] detects two structural paradigms, namely a comprehensive approach that results in a single model or rather a modular approach where more models are executed in parallel. Accordingly, when simulation is applied to a logistic system, the corresponding simulation model can reproduce all the nodes of the supply chain [13] or, instead, can integrate multiple interacting models (one for each node) running according to the paradigms of parallel and/or distributed simulation [3, 4, 14].

Furthermore, when considering methodological underpinnings of simulation approaches, software engineering is inevitably referred given that simulation building blocks are pieces of software components integrated with one another. In particular, Agile Software Development stands out against traditional software engineering approaches especially for complex and nondeterministic software development projects. As well-known Agile Software Development entails a set of methods and methodologies that can be suitably applied in software development when self-organizing and cross-functional groups are involved and when requirements such as adaptivity, reliability, and responsiveness are imposed. Therefore, due to their inherent nature, profitable results can be achieved when agile practices are absorbed into Modeling and Simulation projects. As a matter of facts, a lot of common ground can be found if well-established M&S principles are evaluated against agile practices or vice versa. Indeed, agile software development, likewise simulation, relies on human and technical factors management along the project life cycle. However, it is worth noticing that simulation can provide the groundwork to assess and validate the effects of agile practices and in turn can substantially strengthen its methodological foundations benefiting from advances and best practices in software engineering.

To this end, this research work highlights that some of the agile development theory pillars are, at the same time, driving requirements in M&S projects confirming the idea that when developing a simulation model, best practices can be found out thanks to agile development. In particular, a case study in the field of military logistics is proposed as proof of concept to show that high-quality results can be achieved when agile techniques are applied the development process of a simulation model. The paper is organized as follows: Sect. 2 reports a brief overview on the Agile Software Development state of the art; Sect. 3 is centered on Military Logistics issues and principles and therefore is meant to outline the reference framework the application example (described in Sect. 4) pertains to; Sect. 4 is devoted to show how Agile Methods and Modeling and Simulation can be integrated to develop successful decision support tools in Military Logistics; lastly, findings, future developments, and conclusions are presented.

2 Agile Software Development: A State of the Art

Owing to continuous advances in the ICT sector, software development has as gained an undisputed importance both in the scientific community and in the market. As a matter of facts, customer needs evolve in unpredictable ways becoming increasingly demanding and with growing expectations toward technological products and software applications. It poses serious challenges to researchers and practitioners that are concerned to follow and/or anticipate technological trends.

To this end, since 1968 Software Engineering is committed to provide sound methodological and scientific foundations for software development even if a first attempt dates back to Benington [15]. Indeed, as discussed in Ghezzi [16], software development can turn into a complex, time-consuming, and rather troublesome process. In this context, Agile software Development stands out from other, more rigid and less flexible. Software Engineering approaches as a pattern or rather a mental approach that goes beyond any collection of rules. As well known, It encompasses a set of methods and methodologies for complex and nondeterministic software development projects such as The Dynamic Systems Development Method (DSDM), The extreme programming (XP), and other structured methodologies [17] (i.e., Crystal method, Feature-Driven Development, Lean and Development, Scrum). Agile Methods have gained widespread acceptance within the international scientific community even if there have been more applications in Research and Development, than in Industry. However, on the other side, the study carried out by Begel and Nagappan [18] at Microsoft shows that valuable benefits are perceived and achieved at the highest level of software development. In particular, as discussed in Dyba and Dingsøyr [19], noteworthy benefits include enhanced customers' collaboration, availability of dedicated work processes for defects handling, work revising and engineering, possibility of learning pair programming, ability to think ahead, and outside commonplaces, Nevertheless, Agile Methods have their limitations it is worth being aware of. Indeed, as discussed by Vijayasarathy and Turk [20] worrisome issues include limited support for development activities involving distributed environments and large teams and steep learning curves.

3 Military Logistics

Since the very early design stages, military logistic systems are meant to support the needs of armed forces in critical contexts [21, 22]. For this reason, they require a flexible configuration characterized by a lean command and control chain allowing for different exploitation patterns and screenings along operations.

Lessons learned from past-year experiences in military operations have clearly shown the particular contingents of new supply chain configurations with the necessity of Modeling and Simulation to validate the correct option.

In addition, the fast-changing scenario of each operation has pointed out how difficult it is to define stable and consolidated user requirements for logistic software applications.

Future operations will be characterized by a great degree of variability of the basic parameters. The logistic organization will have to cope with a lower number of assets in the operational theater but with longer delivery routes often located in hostile territory. The nature of the threat will be varying on a wide spectrum of possibilities and conventional delivery means seem to be inadequate for such an environment.

Agility and modularity have to coexist with sustainability as well as with the net-centric nature of the Command and Control structure.

In particular, the logistic tool will have to be characterized by the same degree of reactivity of the supported forces. The reduced number of the assets will oblige to a higher level of efficiency.

New methods are required to understand the level of efficiency for the assets and the value added the Army can bring to the effort is the availability to test and experiment in a Concept and Development Evaluation fashion all the technological and procedural solutions at the earliest stage.

Such solutions may include: Supply chain configuration, organization of clearing operations, high charges for transports tailored to the contingent situation, and forecast of spare parts availability before a malfunctioning becomes critical.

A real challenge will be the transition from a statistic-based prediction to an advanced diagnosis.

The current diagnosis is based on empirical methods greatly depending upon subjective observation. The faults history is never complete and the connection to the symptoms is purely statistical.

Key factors for the change will be the availability of real-time diagnostic tool at the lowest level possible.

From these data through processing and by utilization of synthetic environment, proper sizing and optimization of the logistic chain will be possible.

New technology and competences are required to extend the health monitoring to the minimum component level.

In accordance with the NIILS (Normativa Interforze per il Supporto Logistico Integrato) [23], the Program Management Office (PMO) has great responsibility in the military logistic system definition and carries out organizational and technical management functions during the entire acquisition process, in order to ensure effective operational systems and pre-established goals of costs performances.

Within this context, it is important to have awareness and control during the entire life cycle of the operating system, and ensure the integrated management of all logistical requirements and product data.

As previously discussed in Sect. 1, Modeling and Simulation is an excellent tool in order to support the Program Management Office in the fulfillment of its technical and managerial responsibilities during the entire life cycle of complex systems. Simulation could improve the system design phase and help the PMO to obtain a complete and integrated operating system in order to strengthen monitoring, control, and ensure targeted corrective actions that allow subsidiary and aware cost management (Longo [4]).

Focusing the attention on the ILS (Integrated Logistic Support), the use of the simulation model (that is described in Sect. 4), makes operational analysis efficient and fast, allowing the definition of supportability requirements and the implementation of new standards or parameters modification compared to the already-existing configuration, without any change to the real system.

It could be particularly useful in order to study different operational scenarios, as different mission profiles, specifying for each of them all the operational criticalities in terms of capabilities, mobility, portability, interoperability, and environmental conditions but most of all for the prognostic approach in order to ensure an effective and sustainable system [24].

The multiple functions and features of the simulation model (as described later in Sect. 4) can also support further hard logistics management choices as that of maintenance policy, which defines the distribution of responsibilities for the upkeeps required to maintain or restore the operation of the SP (primary system). Indeed, simulation analysis allows obtaining a greater awareness on multiple aspects of maintenance such as the level of replacement, maintenance locations, operators' skills, in order to minimize downtime, decrease failure rates, strengthen maintenance control, corrective activities, and especially reinforce a prognostic approach. The simulation model even allows gaining feedback of statistical and historical surveys useful to improve the reliability of the system, predict, and contain the extent of losses and damages resulting from a failure by implementing protective actions.

For this reason, the Agile Method for Simulation Models development is able to deal with the complexity of the "Requirement specification" phase, focusing on interactions and not on processes so as to eliminate various levels of "translation" of the user needs and as a consequence limiting the possibility of misunderstanding the set of given specifications. To this end, the Joint use of Agile Methods and Modeling and Simulation Methodologies is envisaged as a promising approach to design, analysis, implementation, and review.

4 LASM—Logistic Analysis Simulation Model

LASM simulation model has been developed as a decision-support tool for monitoring, management, and control of military logistics. As mentioned before, Agile principles have been applied along the simulation model development and implementation so as to address basic requirements such as flexibility, time efficiency, and modularity. For usability purposes LASM is equipped with user-friendly interfaces that provide an easy access to simulation setup, execution, and post-processing. Therefore, front-end interfaces are meant to ensure LASM fully deployment even among nonspecialists users that can avail of such interfaces to:

- configure the supply chain and logistic scenario;
- run the simulation;
- observe, analyze, and export simulation results.

Indeed, the main dialog window provides users with commands to set scenario parameters (number and type of resources, number of supply chain echelons, number of nodes and position, inventory management policies, demand forecasting methodologies, transportation strategies, etc.). Moreover, as shown in Fig. 1, interfaces include even control commands for simulation execution management (start, restart, shutdown, and running) and boolean control for the random number generator (in order to reproduce the same experiment conditions in correspondence of different operational scenarios).

It is worth noticing that the simulation model is parametric so that changing the main parameters different supply chain configurations and operational scenarios can

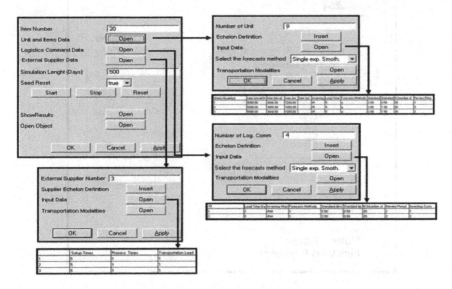

Fig. 1 Example of graphic user interface for the LASM simulator

be obtained. Thus, PMOs can use the simulator for evaluating alternatives and carrying out what-if analyses to find out the solution that best meets the need for adequate forces keeping costs as much as possible low. LAMS ambition is to speed-up decision-making processes, improve decision outcomes (and effectiveness) effectiveness under uncertainty, while minimizing risks and unintended secondary effects.

To tackle uncertainty, LASM captures the dynamical behavior as well as the stochastic nature of Military Supply Chains taking into account the constraints and limitations that are typical of real-word operational scenarios. In this way, it is able to provide a realistic operational picture of the system under investigation as well as feasible solutions (i.e., forces allocated to specific operations) that can be easily implemented in real theaters. Optimal policies in resources allocation are ensured by an optimization tools that has been integrated into the LASM simulation framework so as to take advantage from the joint and combined use of simulation and optimization (as depicted in Fig. 2). On one hand, the optimization module implements several algorithms (i.e., Ants Colony Optimization, Genetic Algorithms, Tabu search) with the aim of optimizing multiple performance measures (i.e., late deliveries, systems breakdowns, orders cancellations, increased

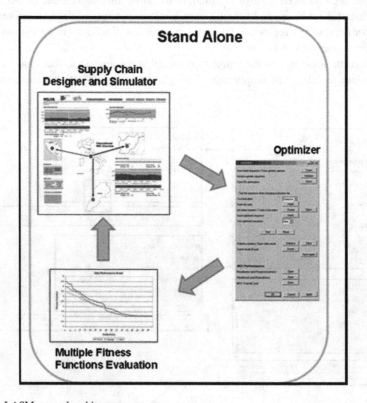

Fig. 2 LASM general architecture

inventories, additional capacities, or unnecessary slack time). On the other hand, the simulation environment can be used as a test bed for evaluating how optimized solutions impact over the whole Military Logistics (before putting hands on and take decisions in the real system) so as to enhance operational/context awareness and minimize unexpected consequences or externalities that could prevent military operations from being successful.

Thus, Responsible Officers using LASM can improve their context awareness, acquire a deeper understanding on the impact of their decision on the overall MSC, and at the same time improve their prediction capabilities.

Furthermore, LASM is meant to support Military Logistic planning leveraging on forces mobilization and allowing for maintenance models able to adapt to changing operational conditions. Moreover, additional capabilities include the possibility to assess various repair patterns, to name a few:

- RMA (Reliability and Maintainability Analysis)
- Analysis of the system operational requirements
- Analysis of training requirements and instructional staff
- RCM (Reliability Centered Maintenance)
- LCCA (Life Cycle Cost Analysis)
- MTA (Maintenance Task Analysis)
- FMECA/FMEA (Failure Mode and Effect Analysis, i.e., the evaluation of the possible failure modes)
- SO (Spare Optimization, analysis of parts and sizing of stocks)
- LORA (Level of Repair Analysis)
- Operator Task Analysis
- Infrastructure and tools analysis
- RAMS (Analysis of Reliability, Availability, Maintainability, and Safety).

Furthermore, LASM use modes include a Fully Federated Operational Mode since it has been designed to comply with the IEEE 1516 interoperability requirements. As a consequence, LASM can be easily integrated into an HLA (High-Level Architecture) federation and work in cooperation with other simulation models.

5 Agile Methods and M&S Principles Along the LASM Life Cycle

Once the LASM potentials and capabilities have been discussed, it is worth focusing the attention on the methodological and implementation effort that has been required to have LASM in its fully operational configuration. First of all, along the development process a strict conformity of M&S methodologies to Agile Software development principles has been noticed. As a matter of facts, being centered on the concept of customer involvement, agile practices meet one of the key requirements in a simulation study: customers can provide meaningful insights and domain knowledge that the modeling and coding phases could substantially take advantage of. As for LASM, it has been achieved involving Academia together with Army that is envisaged to be its main end user. In particular, the development process has highly benefited from a multidisciplinary and cohesive team where much attention has been paid on cross-domain interactions, knowledge exchange, and effective communications. It has been a fundamental premise for capturing the complexity of operational and planning issues behind Military Logistics. Indeed, as previously mentioned, Military Supply Chain involves many interacting components, nonlinearities, and constraints that have been captured within the LAMS simulation framework. From a technical point of view, being much focused on the deployment, quality, reliability, and reuse, both M&S and Agile Development have driven toward an adaptive and iterative development process. As a matter of facts, building a simulation models embeds a software development process and therefore applying Agile methods is quite effortless. It has been also recognized within the scientific community; for instance, Sawyer and Brann [25] show that using the Milestones approach (that is an Agile practice), opportunities for frequent and recurring testing become a natural part of the simulation project life cycle. As for the experience matured during LAMS development, Extreme Programming (XP) has been applied getting good results in terms of software reliability, shorter testing, and debugging times. In particular, the simulation model implementation has highly benefited from pair programming, unit testing, and refactoring that are XP key components. Unit testing and refactoring have fed into simulation verification processes making the simulation model extremely robust and error-free. Further benefits, achieved thanks to Agile Software Development principles and XP include improving the capability of making changes and modifications and of integrating effectively all the software components that build the simulation model up. Moreover, XP has gained the development team approval owing to the positive effect over some software properties such as reusability and extendibility. The whole team has recognized an improved capability of meeting the above requirements.

6   Conclusion

Modeling and Simulation is recognized as a very efficient tool in order to support the military logistics management. It allows dealing with many critical issues at strategic, operational and tactical level, and overcoming all the contextual and political constraints usual of a war theater. Basing on the assumption that the structure of a simulation model consists of a set of software components well integrated each other, the purpose of this paper was to analyze the impact of the Agile Methods and principles over the simulation model development process. In particular, the Extreme Programming method has been applied during the development of a simulation model, LAMS, for decision support in Military Logistics. It has been a profitable experience since the whole development team agrees in recognizing that good results in terms of software reliability, shorter testing, and debugging times have been achieved. Moreover, simulation verification processes have highly benefited from unit testing and refactoring making the simulation model extremely robust and error-free. Further benefits include an improved capability of making changes and modifications, of integrating effectively all the software components that build the simulation model up, and of meeting requirements of extendibility and reuse of software components.

References

1. Prebilic V (2006) Theoretical aspects of military logistics. Defense Secur Anal 22(2), 159–177
2. Perugini D, Lambert D, Sterling L, Pearce A (2002) Agent technologies in logistics. 15th European conference on artificial intelligence July 21–26, Lyon, France
3. Bruzzone A, Garro A, Longo F, Massei M (2014) M&S support for system of systems engineering applications. In: Chap. 8, An agent oriented perspective on system of systems for multiple domains. Wiley, New York, pp 187–217
4. Longo F (2011) Industry, defence and logistics: sustainability of multiple simulation based approaches. Int J Simul Process Model 6, 245–249. ISSN: 1740–2123
5. Slats PA, Bhola B, Evers JJM, Dijkhuizen G (1995) Logistic chain modelling. Eur J Oper Res 87:1–20
6. Yücesan E, Fowler J (2000) Simulation analysis of manufacturing and logistics systems. In: Swamidass P (ed) Encyclopedia of production and manufacturing management. Kluwer Academic, Boston, pp 687–697
7. Banks J (1998) Handbook of simulation: principles, methodology, advances, applications, and practice. Wiley, New York
8. Longo F (2011) Advances of modeling and simulation in supply chain and industry. Simul Transac Soc Model Simul Int 87:651–656
9. Curcio D, Longo F (2009) Inventory and internal logistics management as critical factors affecting the supply chain performances. Int J Simul Process Model 5(4):278–288 (ISSN: 1740-2123)
10. Bruzzone A, Massei M, Longo F, Poggi S, Agresta M, Bartolucci C, Nicoletti L (2014) Human behavior simulation for complex scenarios based on intelligent agents. In: Proceedings of the 47th annual simulation symposium, simulation series, vol 46, pp 71–80, SAN DIEGO: SCS, Orlando, FL, USA (ISBN: 978-1-63266-213-2, ISSN: 0735-9276) 13–14 Apr 2014
11. Longo F (2011) Supply Chain Coordination and Management. Pengzhong Li, Cap. 5, Supply chain management based on modeling and simulation: state of the art and application examples in inventory and warehouse management. InTech—Open Access Publisher, pp 93–144
12. Fujimoto R (1999) Parallel and distributed simulation. In: Proceedings of the 1999 winter simulation conference, IEEE, Piscataway, NJ, pp 122–131
13. Bruzzone AG, Longo F (2014) An application methodology for logistics and transportation scenarios analysis and comparison within the retail supply chain. Eur J Ind Eng 8:112–142. doi:10.1504/EJIE.2014.059352 (ISSN: 1751-5254)
14. Bruzzone AG, Longo F (2013) 3D simulation as training tool in container terminals: the TRAINPORTS simulator. J Manuf Syst 32:85–98 (ISSN: 0278-6125)
15. Benington HD (1956) Production of large computer programs. In: Symposium on advanced computer programs for digital computers, Washington, D.C., Office of Naval Research

16. Ghezzi C, Jazayeri M, Mandrioli D (2003) Ingegneria del Software 2nd Edition Pearson Education, Prentice Hall. Pearson Education, Prentice Hall
17. Agile Manifesto (2015) Manifesto for agile software development. http://agilemanifesto.org/. Accessed Jan 2015
18. Begel A, Nagappan N (2007) Usage and perceptions of agile software development in an industrial context: an exploratory study
19. Dyba T, Dingsøyr T (2008) Empirical studies of agile software development: A systematic review. Inf Softw Technol 50:833–859
20. Vijayasarathy LR, Turk D (2008) Agile software development: a survey of early adopters. J Inf Technol Manage 19(2):1–8
21. La Dottrina Logistica dell'Esercito N. 6623 EI – 4A ed. 2000
22. Manuale S4-G4 dell' Esercito Italiano
23. SGD-018 (2009) NIILS—Normativa Interforze per il Supporto Logistico Integrato (ILS-Integrated Logistic Support) ed.
24. Halfpenny A, Kellet S (2010) pHUMS—Prognostic Health and Usage Monitoring of Military Land Systems. Proceedings of the 6th Australasian Congress on Applied Mechanics 1158–1166
25. Sawyer JT, Brann DM (2009) How to test your models more effectively: applying agile and automated techniques to simulation testing. In: Proceedings of the winter simulation conference (WSC '09). Winter Simulation Conference 968–978. ISBN: 978-1-4244-5771-7
26. Rao KN, Naidu GK, Chakka P (2011) A Study of the Agile software development methods, applicability and implications in industry. Int J Softw Eng Appl 5(2)

Software Characteristics for Program Forza NEC Main Systems

Angelo Ing. Gervasio

Abstract FORZA NEC is the main program for the digitization of the Italian Defense Land component, delivering the pillars of the digitized Italian land and amphibious Forces. The object of the program encompasses the development and procurement of high-technology system capabilities, including C2, Communications, ISTAR, M&S, UAV/UGV. The hardware and software components of Forza NEC systems should be networked and able to operate in unstructured, unsafe environment characterized by wide-range temperature, various terrains, and all weather conditions. This is particularly true for software components, which is playing a leading role in the modern military systems, determining its functionalities and networking features. Therefore, software for military systems must be characterized by the following: high reliability, high level of security, maintainability, user-friendly, standardization, modularity, reusing, upgradability.

1 Introduction

New operational scenarios, integrated joint, and multinational context and constant technology progress drive the development of military digitized Forces in the frame of the network-enabled capability (NEC) information superiority, situation awareness concept.

The core of the NEC is the network that interconnects all the players and platforms to gain the value of their interactivity and achieve desired effects consistent with the operational objectives.

General Secretariat of Defence and National Armaments Directorate—Land Systems Procurement Agency

A. Ing. Gervasio (✉)
Ministry of Defense, Rome, Italy
e-mail: angelo.gervasio@esercito.difesa.it

© Springer International Publishing Switzerland 2016   145
P. Ciancarini et al. (eds.), *Proceedings of 4th International Conference in Software Engineering for Defence Applications*, Advances in Intelligent Systems and Computing 422, DOI 10.1007/978-3-319-27896-4_13

The complexity of the future war fighting environment will require that information be securely and reliably transmitted over dynamic and potentially unreliable virtual and physical networks. Data from a wide range of systems and sensors need to be fused, analyzed, and summarized to help support the rapid and effective decision making.

Creating software to manage this modern C2 functionality provides a number of significant computer science challenges and needs improvements in software development productivity and quality.

These challenges mirror in the Program Forza NEC, which is the tool to provide Italian land assets with "state-of-the art" NEC capabilities.

This paper gives a short background of the Program Forza NEC and illustrates the main characteristics in terms of structure and main systems.

Then follows the description of the main projects of the program that uses extensive software component and highlights their main characteristics in light of the fundamental network and digitization requirement.

2 Background

Italian Army started, in early 2000s, the transition from stand-alone/analog asset functionality to networked nodes of a homogeneous network.

This decision was taken on the wave of the technological progress and in order to build a modern Force, able to effectively operate in the future joint and multinational operational environment.

The transition initiative took inspiration from the network-centric warfare (NCW) concept, developed in the USA in the late 1990s. At that time, it became clear that this process would not be easy to achieve and would imply a complete transformation not only of the Army but of the entire Italian Defense Land Forces.

This was the reason why the Italian Joint Military Staff approved a major procurement program called Forza NEC, where the acronym "NEC" stands for network-enabled capability, following the UK and NATO evolution of the US NCW concept.

The initial scope of the program was the realization of three fully digitized Brigades and a land–amphibious component. The long foreseen time frame (over 30 years) suggested to introduce preliminary steps, in order to get operational spin-offs. Unfortunately, the recent budget cut, the continue downsizing of the military personnel and the evolution of the initial requirement either operational or technical, limited the future phases of the program, which for the moment will end in 2018–2020 with the current concept development and experimentation (CD&E) phase.

The initial requirement will be completed by another program, currently under funding evaluation, linked with Forza NEC.

3   Main Characteristics of the Program

Forza NEC follows these driving concepts:

- Evolution throughout production: Technological and operational evolution should apply during the production phase;
- Transforming while operating: New systems should be ready to deploy without affecting running operations;
- Capability approach: Architectures, systems, and platforms contribute to realize specific capabilities (Fig. 1).

The program adopts a capability-based planning (CBP) approach and operational requirements analysis in order to identify operationally effective capabilities (Fig. 2).

A "virtuous cycle" has been adopted in order to meet the requirements while maintaining coherence with the technological and the operational evolution. The virtuous cycle consists of the following steps:

- modeling and simulation (M&S): to check technical compliance of the system/platform under test;
- field testing: to verify its operational validity;
- deployment: to insert system/platform in the real force (Fig. 3).

Fig. 1 Forza NEC driving concepts

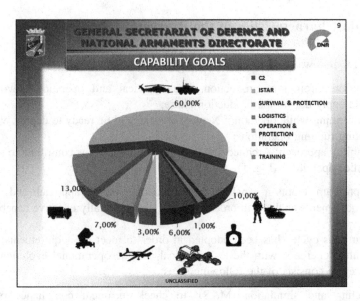

Fig. 2 Capability goals

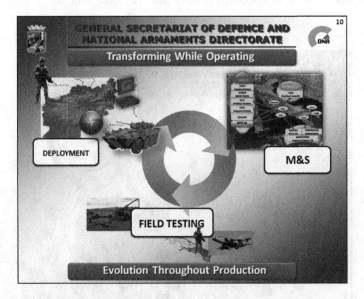

Fig. 3 The virtuous cycle

4 Organization and Procurement Aspects

Forza NEC is a complex program divided into 3 phases for the realization of 35 main projects (Fig. 4).

At the moment, the program is in the CD&E phase that is the prototyping step. The goal of the CD&E is to deliver small series of systems/platforms ready for mass production, with an high degree of confidence to fulfill the operational requirements.

The procurement organization of the Program Forza NEC involves operational and technical–administrative institutions. In particular, the Italian Ministry of Defense addressed the Italian Minister of Economic Development in order to get special funding, dedicated to promote the national industries. Therefore, the National Land Armament Directorate, responsible for the contract, selected Finmeccanica company SELEX ES, former SELEX Sistemi Integrati, as prime contractor (Fig. 5).

SELEX ES interacts with the ministry of Defense through a Program Directorate Forza NEC, constituted inside the General Secretariat of Defense, joining the Procurement Agency, the Operational Staffs, and the industry.

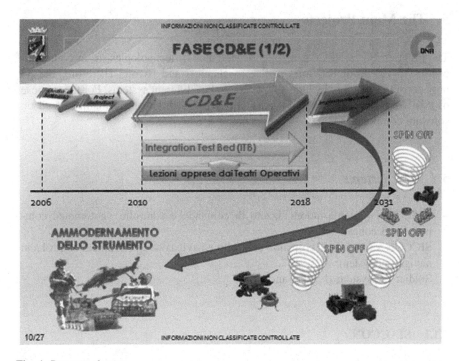

Fig. 4 Program phases

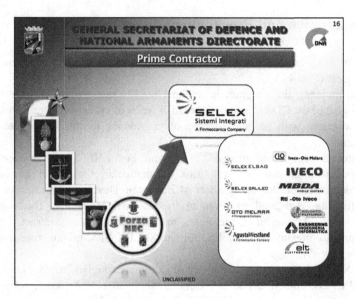

Fig. 5 Procurement structure

5   The Main Projects

The Program Forza NEC spans over a wide area of technologies, from communications to electro-optics, informatics, mechanics, and so on. Therefore, the 35 projects of the program include a huge amount of different systems. For the purpose of this work, this paper will consider only the systems that are mainly based on software components (Fig. 6).

5.1   C2 Systems

– SIACCON (sistema automatizzato di comando e controllo), automated command and control system;
– SICCONA (sistema di comando controllo e navigazione), command control and navigation system;
– Soldier C2 system (Figs. 7 and 8).

5.1.1   SIACCON

SIACCON is the main C2 system for battalion/regiment and above, supporting commanders, and their staffs in the following activities:

Fig. 6 Forza NEC main projects

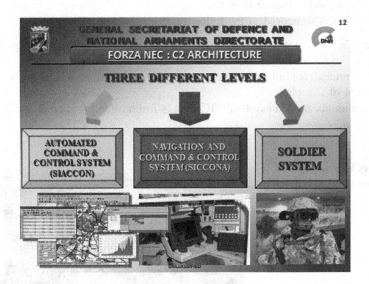

Fig. 7 C2 structure

- tactical situation analysis and monitoring;
- planning and order generation;
- management of the different staff elements;
- support to military operations;
- information and data exchange with other joint and multinational equivalent systems.

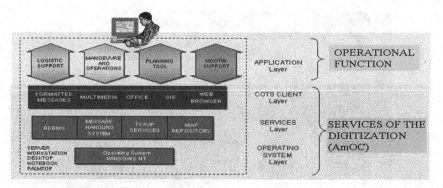

Fig. 8 SIACCON structure

5.1.2 SICCONA

SICCONA is the second level of the C2 chain and is built for combat and transport vehicles and for most of the tactical weapon and sensor systems. SICCONA integrates single platforms in a common network, gathering position, data, information, orders, and logistic situation (Fig. 9).

SICCONA is composed of the following:

- a vehicle integration subsystem (SIV) that interfaces sensors and other platform assets;
- the command control and digitization (C2D) subsystem, which is the node terminal of that platform;
- the communication subsystem (SIC) that is responsible for internetworking the platform.

Fig. 9 SICCONA main subsistems C2D and SIV

SICCONA provides the crew with situational awareness of current operations, including the following:

- location of friends and enemies, with the eventual sensor;
- platform and unit status;
- operational overlays;
- horizontal and vertical information exchange.

5.1.3 Soldier C2 System

Soldier C2 System uses a wearable computer that supports data processing activities, including sensor data collection, management, and dissemination (Fig. 10).

The soldier C2 system features modern wired and wireless (Bluetooth) standard interfaces to connect with a huge types of radio, electro-optic, physiologic, and other kind of sensors.

The software runs over different portable computers, including Windows CE and Android.

Man machine interface is provided by a built-in or separate touch screen display, available in two sizes: 3.5″ for the soldier and 6″ for the commander.

The position of the soldier is acquired by a GPS receiver, embedded in the radio or external, managed by the C2 software, that determines the position on the map and provides the overall situation updates.

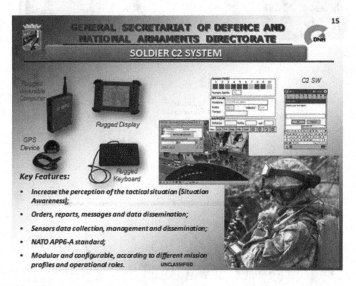

Fig. 10 Soldier C2 system

5.1.4 Integrated Test Bed (ITB)

ITB is a network of Army, Air, and Navy M&S physical sites, located in different places of Italy and connected through the national military residential fiber optic network (Fig. 11).

All sites share a common synthetic environment (CSE), represented by a geographical picture where different kinds of virtual, simulated, and real assets play predetermined or runtime joint or isolated missions.

Each center is equipped by the following subsystems:

– scenario generation and animation;
– online trial monitoring;

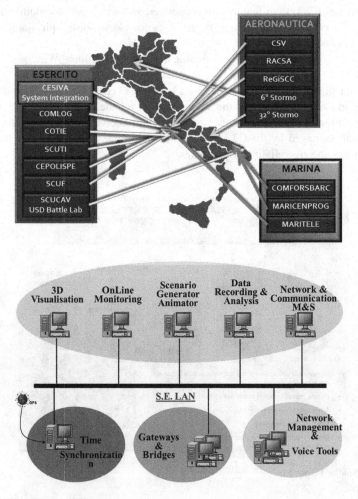

Fig. 11 ITB sites and CSE main components

- data recording and analysis;
- communication assets simulation;
- stub generation;
- C2 and other systems gateway.

5.1.5 Software Defined Radio (SDR)

SDR represents the railroad for tactical communication assets. Italian Ministry of Defense invested a lot or money and efforts on this technology, achieving excellent results. Tactical 4 channel vehicular and handheld radio are the main equipment included in Forza NEC (Fig. 12).

The SDR 4Channel features 4 independent transceivers span from 2 MHz to 2 GHz, a NETSEC central unit for TRANSEC and COMSEC, and a power supply. The size of the radio is form and fit with a dual SINCGARS vehicle installation, facilitating quick and easy transition.

The 4Channel SDR can host a huge variety of waveforms, with different kind of modulations, including wideband data MANET (Fig. 13).

The handheld radio is a multiband SDR for tactical mobile communications that provides secure voice and data services.

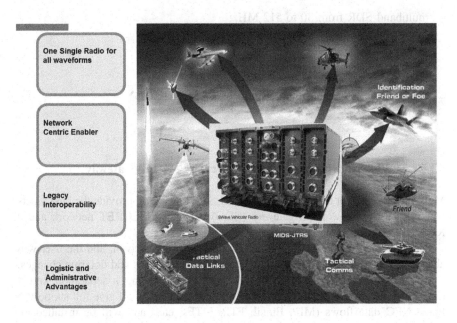

Fig. 12 4 channel SDR for vehicles

Fig. 13 Hand held soldier
SDR and its vehicular adapter

It comes in a simple handheld version for the soldier, but the same handheld fits in a cradle for vehicle installation. The handheld SDR main characteristics are as follows:

– multiband SDR from 30 to 512 MHz;
– proprietary SCA operating environment;
– SCA 2.2.2 compliant;
– narrowband and wideband waveforms.
– embedded GPS receiver;
– embedded 5 W power amplifier;
– encryption capability.

5.1.6 Multiple Independent Levels of Security (MILS) Gateway

Multiple independent levels of security (MILS) gateway will provide the land assets with a secure cross-domain gateway (CDG) between a SECRET network and a RESTRICTED one (Fig. 14).

The information flow will be bidirectional and policy driven, and the user will able to modify the policy according to the operational needs and doctrine changes.

The MILS gateway paradigm will be enforced adopting a separation kernel (SK) evaluated by US NSA (Common Criteria EAL 6+). It will give full support to Forza NEC data flows (MIP, Email, FTP, MTFs, etc.) and will be installed on vehicles and command posts (shelters, tents).

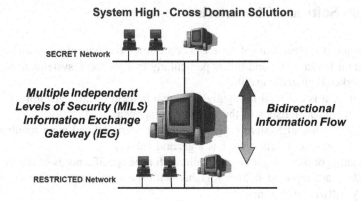

Fig. 14 MILS structure

6 The Importance of the Software for the Program Forza NEC

As highlighted before, software is a primary component of Forza NEC main systems, platforms, and architectures, giving a big contribution to tactical operations success. Therefore, it is of paramount importance to focus on reliable, effective, efficient, interoperable software, while improving the affordability and quality of this increasingly crucial element of modern weapon systems.

To this extent, it is necessary to alter the common perspective, processes, and capabilities in order to maintain the necessary quality and performances, and avoid the increasing costs associated with software development, modernization, and sustainment.

Advanced software systems and embedded software technologies become the brains behind modern warfare, and the amount of software code in modern war fighting systems, including those of the Program Forza NEC, has increased significantly over the last years.

At the time of the program starting phase was adopted the MIL STD 498 methodology, in order to guarantee configuration control, reusability, and quality of software. Over the past years, the huge amount of documents and execution steps foreseen by the MIL STD 498 slow the delivering of systems and overloaded the effectiveness of the procurement process.

A common milestone of the program, for the delivering of a piece of software or equipment, includes the elaboration of different documents during the phases of preliminary design, requirement review, and critical design, just to start with the production.

For this reason, particularly for software development that requires close interaction between user and producer, the Army staff is introducing new development techniques based on agile software production methodology.

7 Main Software Characteristics

At this point, it is clear that software is a cornerstone component for main assets of the Program Forza NEC, and this is particularly true for those systems that realize the networked/digitized capability.

Therefore, it is essential to highlight the requirement of this capability, in order to derive the characteristics of the software component.

US Future Combat System program, today aborted, defined main requirements for the networked capability that has a general validity.

Summaries of these requirements, tailored for the specific needs of the Program Forza NEC, are reported below, together with the consequent influence on the necessary software component:

- collect, display, and disseminate a common operating picture (COP) and environment information which support dynamic mission planning/rehearsal, using also virtual decision-making capabilities.
  This requirement means that the software system must realize and maintain a real-time, easy-to-understand, and accurate COP. Therefore, the volume of information distributed throughout the battlefield sensors and systems network must be rapidly and accurately integrated, then analyzed, and organized to support military decisions. Furthermore, it is important to have online M&S capabilities, to support and speed up decision making;
- enable battle command, on the move, supported by C4ISR architecture for continuous estimate of the situation and share the COP to enable visualization and dissemination of tactical scheme.
  To meet this functional requirement, the system software must have the ability to move command securely from one platform (node) and/or commander to another. This type of command requires that system software supports the ability to deliver orders when one or more of the participants are moving. This function would also have to be tightly integrated with the physical C2 network;
- contain networked information system that enables commanders to effectively lead during dynamically changing operations anywhere on the battlefield. This includes the following tasks:

  - maintain situational understanding at all times;
  - collaborate with subject matter experts, subordinate commanders, and staff;
  - identify schemes of maneuver, opportunities, decisive points, terrain and weather updates, vulnerabilities, solutions through collaborative planning, rehearsal, and simulation;
  - make reasoned decisions, based on information available. The commander will be able to leverage intelligent agents in his information systems to assist him in filtering through the vast amount of information;

- communicate orders, intent and operational information, either text or graphics;
- synchronize maneuver, fires and reconnaissance, surveillance, targeting, and acquisition (RSTA);
  To meet this requirement, the C2 software and supporting ISR resources must be able to rapidly and accurately acquire and fuse mission-relevant data, then assist in analyzing and summarizing the data, and finally help to support command decisions;

– leverage real-time collaborative planning support tools to achieve knowledge-based course(s) of action development.
  For this reason, systems must be mobile, fully interoperable in the joint multinational, and interagency environment. This poses a significant technical challenge in the area of decision support and security, through different levels of classification;
– provide information and C2 systems with digital mapping tool for high terrain resolution to enable C2 of small unit tactical action in close, complex terrain, virtual rehearsals, and terrain analysis. Also allows visualization of inside buildings and subterranean dimension.
  This implies availability of three-dimensional (3D) visualization to the C2 system concept and requires the software to perform very complex data analysis and calculation;
– enable C2 needed to synchronize fire, maneuver, and RSTA to close with and destroy the enemy. Coordination and synchronization with a decentralized C2 needs temporal requirements and constraints for all C2 functions. This is a critical constrain for software development, particularly in a networked environment, where web and publish–subscribe technologies seem to underestimate the real-time quality of service parameter.

To satisfy the requirements analyzed above, the software development needs to adopt scalability, mobility, and security criteria.

The emerging Forza NEC concept of C2 activities will no longer be performed in a centralized manner, but over a dynamic network of moving vehicles, systems, and platforms and will depend on a vast array of sensors to gather data from the battlefield.

This new C2 network will be created in an ad hoc fashion, with nodes entering and leaving the network at unpredictable times.

The C2 system must be highly reliable and highly secure. The battlefield sensor information, vital to C2, will be broadcast from different locations.

This proposed C2 network must be able to process data rapidly and deliver the right information to the right locations and people at the right time.

Therefore, as highlighted above, this provides the following main software challenges:

- distributed computing over unreliable, ad hoc, dynamic, and inhomogeneous physical network;
- fault tolerance over a system in which, at any given time, it is unclear what nodes are available within the network;
- network security and accessibility for immediate access to the network, preventing adversaries from accessing or corrupting it;
- data fusion, from different systems and sensors;
- information analysis and summary of huge amounts of data on the basis of user needs;
- decision support, through a network capable of supporting C2 decision making;
- software development improvements to reduce the complexity and risk in creating the proposed system.

8 Conclusions

Technology evolution led innovation in every field of the human activity, including military operations. In particular, complex functionalities and performances rely more and more upon software component that pervades every system and erodes continuously hardware role.

Network technology invaded military doctrine on the basis of the NCW concept that introduces situation awareness and information superiority as essential capabilities for success in modern military operations.

On this basis, Italian Ministry of Defense started the transformation of its land assets through the activation of the Program Forza NEC that, though has been limited in dimension, maintain its scope on the networked and digitized main land assets realization.

This paper examined the evolution and main characteristics of this important program, focusing on the systems that require extensive software development:

- C2 SIACCON, SICCONA, and Soldier C2 system;
- ITB;
- SDR;
- MILS gateway.

Then, the work listed the main requirements of the Forza NEC networked capability that is mainly based on the above-mentioned systems, highlighting the role of the software component and extracting its main characteristics.

From this analysis, it is possible to conclude that software is particularly important in military systems, and its characteristics are peculiar and dedicated to fulfill operational requirement.

Therefore, it is dangerous to consider software for military systems at the same way as software for equivalent commercial systems, mainly because of the heavier environmental conditions of the operational scenarios, real-time functionalities, robustness, and very high reliability.

Configuration, cost, and quality control of this peculiar kind of software are a challenge and require dedicated approach and expertise.

To this regard, the Program Forza NEC adopted the methodology described in the MIL STD 498 that proved to be complete but lack of effectiveness.

In particular, this standard defines numerous steps and document elaboration in order to guarantee configuration and quality control that require huge amount of time for discussion, requirement analysis, and design definition, slowing considerably the realization time.

Sometimes, the software realized fulfills perfectly configuration and quality criteria, but did not meet the requirement changings.

From this experience, the Program Forza NEC is experimenting different techniques for software development, using modern agile software methodologies that can better follow user's changing requirements, shortening delivering time.

To this regard, the Army staff started the development of a new version of its land C2 system, called land C2 evolution (LC2EVO) using the SCRUM agile methodology, based on the continuous realization of three-week software sprints, led by user representative.

The challenge in this case is the adoption of a suitable configuration and quality control that guarantees the achievement of common basic software characteristics, already mentioned, such as real time, reliability, cost control, and integration with legacy and future systems.

Agile Plus New Army Diffused and Shared Leadership

Stella Gazzerro, Antonio Francesco Muschitiello and Carlo Pasqui

Abstract The Integrated Development Team#4 was set up with the objective of developing the functional area service related to the joint intelligence, surveillance, and reconnaissance, which is a peculiar competence of the Italian Army ISTAR-EW Brigade. Based on the Agile Methodology, the close communication and the continuous interaction with the client enabled the team to become familiar with the complex concepts related to the IPB preparation Process. The "C3—Command, Control, and Communication," a pillar in the military art of command, is provided by the IBM Jazz Suite, which is the core of the collaborative approach founded on Agile methodology and supporting: the speed of the participants, the coordination of their effort, and perhaps, most importantly, the communication between all team members even though on geographically distributed areas. Every team member can know which is the team's shared goal and the work progress at every time of the day, to monitor the cost and workload optimization. To sum up, the team-centric nature of this method as well as the peculiar military tendency to focus on group dynamics ensured to build a close-knit team, willing to tackle the challenges that will be arising throughout the next few sprints.

1 Introduction

On January 2015, Italian Army Head Quarter (SME—Stato Maggiore dell'Esercito) accepted the hard task to develop a combat component of the Italian Army for the continuous improvement of LC2EVO System (Land Command, Control

S. Gazzerro (✉)
Profesia Srl, Turin, Italy
e-mail: stella.gazzerro@profesia.it

A.F. Muschitiello · C. Pasqui
Italian Army, Rome, Italy
e-mail: antoniofrancesco.muschitiello@esercito.difesa.it

C. Pasqui
e-mail: carlo.pasqui@esercito.difesa.it

© Springer International Publishing Switzerland 2016   163
P. Ciancarini et al. (eds.), *Proceedings of 4th International Conference in Software Engineering for Defence Applications*, Advances in Intelligent Systems and Computing 422, DOI 10.1007/978-3-319-27896-4_14

Evolution). So, Italian Army HQ staff decided to increase the challenge of introducing it in a program oriented to customize Agile Scrum methodology to feature Italian Army software development lifecycle [1].

The production process of Agile software in the acceptance of Italian Army Agile [2] has allowed to analyze and define the user needs in natural language through the standard template of user stories materialized in quick releases and incremental that led you to a growing trust in "Italian Army Agile" approach.

The method still today is always subject to verification cycle and continuous improvement, and it has proved easy to apply to the following:

- complex problems of Scrum of Scrums,
- team geographically distributed,
- complex scenarios of military doctrine.

It provided support to issues both ordinary and extraordinary types through a combination of methodology and support tools to the process such as the IBM Rational Jazz platform [3].

2 Foreword—The MAJIIC2 Program and the Need of a New Development Methodology

The Intelligence R&D MAJIIC2 (*Multi-Int All-source Joint ISR* (Intelligence, Surveillance, and Reconnaissance) *Interoperability Coalition*) [4] five-year program born on the outcomes of the CAESAR (Coalition Aerial Surveillance and Reconnaissance), CAESAR Extension, and MAJIIC programs, with the aim of making interoperable the new sensor typologies (electro-optical, video, infrared, ESM) with the new information sources. This program also aims at developing a common architecture for data exchange, analysis, and presentation common criteria in order to speed up the JISR processes and the employment operational concepts, as well as techniques, tactics, and procedures allowing interoperability both in the case of multinational coalition and NATO (*North Atlantic Treaty Organization*) operations.

In order to achieve these objectives, the coalition devised a platform made up of the following components:

- CSD (*Common Shared Dataserver*) [4]: This database is meant to carry out a fusion activity, since it contains all ISR data and hosts the synchronization and interoperability activity of the participating nations.
- IRM&CM [4] (*Information Requirement Management and Collection Management*) tool: It represents the C2 operations since it relates to the task management function of the search assets.
- *Exploitation tool*: This tool is meant to exploit the information contained in the CSD. Considering the manifold nature of the CSD data, the exploitation tools are designed to optimize the management of a specific product typology, usually

reproducing the differentiation of the intelligence disciplines, with a preference for the IMINT component—which was extensively developed in the previous programs (CAESAR [5] and MAJIIC)—and the HUMINT (HUMan INTelligence) component, which stands out as a real breakthrough that enabled the multi-intelligence feature of the program.

- Proprietary services allow the correct interfacing of the various tool typologies, e.g., the transmission of the intelligence requests (*Requests for Information*) and the task management relating to the search C2 assets.

The founding nations were ITA, CAN, USA, DEU, FRA, NOR, GBR, with the addition of the NCIA (responsible for system integration and technical management). Later on, also NLD and ESP joined in. On a national level, in 2007 a Programme Directorate (PD) was created under the guidance of the Defense General Staff, with the purpose of setting up a committee working as a catalyst for the national acquired knowledge and experiences. That facilitated the integration into the Defense C4ISTAR architecture and allowed to take care of the implementation of a comprehensive product for the management and processing of the ISR national and coalition data, on a tactical as well as on an operational level.

At the end of the program, Italy developed a national CSD version and related services for the correct functioning, and the ICISRC (*Italian Common ISR Capability*) *exploitation tool*, which was set up to meet the IMINT priority needs. In this regard, every single Armed Force provided its own personnel within the *Subject Matter Expert* to define the requirements and take part in the program's key events, i.e., *Working Groups* consisting in quarterly meetings aimed at discussing the platform implementation progress. As an example, it may be worth mentioning the IRM&CM functionality or the integration of the HUMINT disciplines as well as the exercises of the MAJEX (MAJIIC2 Exercise) and the Unified Vision series. Such exercises are meant to technically assess the level of interoperability with respect to the achievements of every single nation, while from an operational point of view, they are meant to assess procedures, techniques, and tactics for a fruitful ISR data exchange among the 9 NATO participating nations. The point of contact is the RISTA-EW Brigade, whose personnel was constantly involved in the program development, taking part in every single event with the same contact persons.

In this context, it was agreed that the classical planning paradigms should be used, with particular reference to the iteration planning method, based on the following phases:

1. planning,
2. requirements analysis,
3. project,
4. implementation,
5. trials,
6. assessment.

In particular, the first two phases are carried out throughout the sub-working groups, where the operational needs (*Operational WG*) and the technical requirements (*Architectural and Technical WG*) are defined.

The stages of design and implementation were performed by the industrial sector without the presence of military personnel.

The validation activities of the useful product for the purpose of testing are carried out following the evaluation test phase done directly during military exercise which has granted approval for the release of the applications developed (CSD and its tool of exploitation National ICISRC).

This paradigm of software development has proved woefully inadequate to meet the needs of the FA, hindering control of what produced correspond with user request and making it difficult and occasional, the end-user involvement by preventing the customer to correct the development distortions. Evidence of this case emerged only after one year and in the final stage through the use of the products by qualified personnel during military exercise session.

3   The LC2EVO Opportunity

At the end of 2014, the opportunity of starting up the FAS JISR within the LC2EVO was immediately seized by the Network Infrastructures Office since the development of an agile methodology would allow to integrate the national achievements at a joint level (in compliance with the NATO STANAG of the MAJIIC2 program) with what each Armed Force was supposed to develop on its own to meet the peculiar needs arising from its own "distinctive features."

Just as an example, the Army personnel is constantly in touch with the civilian population in the operational theatres, and this is the reason why, when compared to the navy or the air force, it is marked by a much stronger need for implementing further intelligence sources, besides the IMINT (IMagery INTelligence), e.g., the HUMINT, which is a human source. In order to bridge these gaps while meeting at the same time the operational needs, the Army had to regularly integrate the CSD and the ICISRC, with a view to developing the following:

- an IRM&CM tool, as highlighted in the ICMT software (ISR Collection and Management Tool) developed by Norway within the MAJIIC2 program. It became a NATO reference point for the IRM functions (information requirement management) and CM (collection management) to support the CCIRM cell through the RFI information management and the tasking of the research assets;
- a tool for link analysis made up of the I2 Analyst's Notebook, global leader in the application of link analysis;
- a tool to meet the HUMINT's needs, as highlighted in the HOTRA software (HUMINT *On-line Tasking and Reporting*—Allied), which was developed in the USA within the MAJIIC2 program for the workflow management leading to the production of HR (HUmint *reports*);

– a tool for semantic analysis and open source informative correlation, enabling the analysis both of internal documents and of open source materials to support the link analysis activity.

From a logical point of view related to the use, the program MAJIIC2 had already developed the software necessary for the activities of conducting operations, which now is by definition joint, remained to be useful to develop the planning phase characteristic of each Armed Force, and it leads in a different way the consequences of these different distinctive features. In this context, the opportunity given dall'LC2EVO and Agile methodology potentially allowed to coping with all the problems:

– Fully involving this person in the development from the very beginning of the project life;
– Capitalizing on the wealth of knowledge and relationships already established with the staff of the Brigade RISTA-EW MAJIIC2 under the program;
– Checking with cyclicity monthly and annual not more developed than the compliance with the requirement agreed from time to time with the Product Owner (PO);
– Contracting reduces the time and development costs also through the reuse of code and user stories appropriately specialized and refined for the problems to pool.

4 The Onset of the ITD4

The development team 4 was formed for the development of FAS JISR within dell'LC2EVO, with input from the Product Owner represented by Brigade RISTA-EW, and it was agreed in achieving the development of digitized Intelligence Preparation of the Battlefield or IPB which is the process allowing you to create a series of products whose combination leads to identify the essential elements of planning and making predictions about the activities of the opponent.

The choice of the theme to be developed is this motivation in the fact that IPB is the sum of a set of functions to achieve consistent car in a limited number of sprints and optimal developers to become familiar with the method and foster trust by the Product Owner to the team and the method itself.

The Global Strategic Product Owner has made the team 4 choosing its members on the basis of acquired skills and participation in the development of intelligence-sharing program called MAJIIC2, so as to exploit over the synergistic action between the parties already developed earlier and the considerable heritage of relationships and knowledge previously between the various direct personnel by the very high professional quality of the Brigade RISTA-EW.

In this regard, the Brigade's conceptual contribution was key.

The first problem was to choose the best reference cartography to meet the military needs. Therefore, in the first sprint, an informative user story was included. It consisted in a visit of some representatives of the team at the Military Geographic Institute (cartographer, *Scrum Master*, *Deputy Scrum Master*, and *PO*). To this purpose, just consider that the choice of developing this software was beforehand bound to an internal directive on the use of the *software Geographic Information System* (GIS) packages, which identified the ESRI ArcGIS as the Army reference product.

During such visit, it came to light that the Institute was committed to merging the cartographic data contained in civilian and military databases countrywide, while the Sardinia region was ready, having collected and burned all vector data, which were perfectly readable by the ArcGIS.

This is why the Sardinia cartography resulted the most suitable for development. As for the international cartography concerning the operational theaters where the national troops are deployed, throughout the visit it came to light that an international contract had been recently signed in order to share the vector data produced by the cartographic support cells of the various contingents.

As a result, considering that the process is already underway in the Army for vector data systematic provision and sharing in the near future, the digitized IPB was launched.

As for the development, as we already mentioned above, it was decided to adhere to the PO's chronological order. The objective is to deliver a first mock-up product on the occasion of the Army annual intelligence exercise which is going to take place in May. This exercise is called "Galileo" and represents an important achievement of IPB first steps that consists in the definition of the operational environment and the productions of some overlays.

5   The Progress of the ITD6

The complexity level grew when joined to the development of the FAS JFIRES & Targeting.

This choice stems from the successful integration experimentation, which was carried out in October 2014 throughout the annual artillery exercise in Sardinia, named "Shardana."

On such occasion, there was an embryonic integration of the Army Artillery Command Fire Information System with the MAJIIC2 software.

Basically, the SIF could produce a polygonal and then make a *query* on the intelligence database concerning all data matching the polygonal area. On that very occasion, it was demonstrated that a bridge between artillery fire (hence the fire information system) and the intelligence *database* was viable to cut the analysis at the base of the artillery automated operations.

In the end, the assumption that engineers and artillery, which are the *"Combat Support"* branches, are the intelligence-privileged users in current conflicts proved to be true; hence, it was possible to integrate the two environments by means of the information systems. The overlapping of team 6 and 4 definitely concerns the operational environment, its cartographic representation, and its related products.

Needles to say that the reuse of part of the 4 team written code from an economic viewpoint is considered convenient, both in terms of times and of costs. On the other hand, it would have been advisable to anticipate the implementation of some of the products that were scheduled to be produced in the 3° IPB step by the Artillery "twin" team, just as an example.

6   The Launch of the ITD7

As already mentioned, among the intelligence-privileged users, there are the artillery and the corps of engineers. The artillery's role consists in the kinetic or non-kinetic implementation of the offensive activities as they emerge from the situational framework ± intelligence. On the other hand, the corps of engineers' role consists mainly in force protection, in choosing the most effective tactical strategies to neutralize the systematic use of the enemy's suicide attacks and remote/non-remote Countered IEDs (*Counter Improvised Explosive Devices*), which to date represent the enemy's most effective and convenient tactical methodology. Basically, the idea is to neutralize the tactical advantage of the enemy, who plans and acts fast thanks to the current *franchising* insurgency and terror network. The objective is to carry out timely attacks with the help of the intelligence early alerts that are mostly effective insofar as they show the enemy's vulnerable gaps beforehand, which are made up of material and non-material elements, critical requirements with respect to the enemy capability of carrying out offensive activities.

Introducing Military Engineering and Counter IED staff complete the refining process of IPB through the development and extension of information data from all areas of expertise involved in the Operation Order.

7   The Human Factor Perspective: Capturing the *User Needs* on the Field

The Agile software development approach is team-centric by definition, and this is why the human factor is a key. The teams are made up of a military/civilian mixed module to benefit from the "lateral thinking" effect and ensure the permanent presence of the Army personnel that may have a leading role in the complex institutional and military social relationships.

At first glance, this personnel "jumble" may seem to be a little peculiar, considering the rigid military mindset, which typically matches a hierarchical organization with respect to the creative attitude and less hierarchic organization of the civilian software developers.

Nevertheless, the deployment in complex operational theaters with a close cooperation with the civilian population of different cultures brought about a radical change in the military attitude toward the leadership, meant as "diffused" and inclusive leadership, open to diversity and goal-oriented rather than fussing over procedures.

This new approach implies the development of valuable listening and empathy skills, which proved to be perfectly in line with the "democratic" interaction of the *agile* modality.

To sum up, we can state that the *agile* methodology led the civilians to stick to deadlines, time schedules, and sort of red type which the military personnel is very much familiar with. On the other hand, the military developed a new behavioral attitude that makes their relationship with the civilians much smoother.

As a result, this valuable mix proved to be very successful over the last 4 months: In very short times, all teams proved to be able to closely cooperate and join all agile ritual players altogether, such as the *Product Owner*, the *Scrum Master*, and the *Stakeholder*, thus building a virtuous cycle of trust and common goal sharing, which is key to overcoming the difficulties and setbacks.

Capturing "*user needs*" became much easier in this framework of social relationships and enthusiastic cooperation. To this purpose, a number of communication tools were devised, such as social groups and *streaming* videos, which made real time exchange of opinions, suggestions, and ideas incredibly easier. This was of great help for the development perspective and for the solution of any misunderstandings of setback caused by a poor organizational structure.

In addition to these informal tools to enhance group cohesiveness, there is also a formal tool, the JAZZ platform, to check whether the progress of the group's work actually matches what had been planned or the fair and balanced workload.

8 The Military Doctrine Translated into Engineering: Methods and Support *Tools*

8.1 The JISR Case

The ITD4 implements the JISR FAS, marked by a complex military doctrine with well defined and standardized processes and procedures, deriving from real and true-to-life situations experienced in the operational theaters by the personnel in charge and through coalition or NATO standards.

The JISR environment is operational and is supposed to undergo periodic updates aimed at refining and specifying the outcomes of the IPB preparation process.

After adapting the *agile* method to the quality and security needs typical of the C2, the FAS JISR implementation brings about a new challenge for the working group, which is the evaluation of the possibility of adopting and applying the *Agile* method in a context marked by the following:

- Experimentation, so extremely adaptive
- High complexity of the doctrinal topics

Starting from the principles and the practices as they emerge in the literature, the attention focused on the research of the tools that could lead the whole team to achieve the balance point between the doctrine and Engineering and goes beyond the limits of the traditional method of software production, thus reducing the patent collision between industry and the Army. In light of this interaction, all issues and complex topics, typical of the IPB process preparation, can be correctly understood.

In this framework, the natural language of the *user stories* helped the users define, describe his needs, and face complexities whenever needed, outlining the existing relationships throughout the wide range of topic analysis, also considering the activities related to other FAS, such as the Military Engineering and CIED and the Artillery.

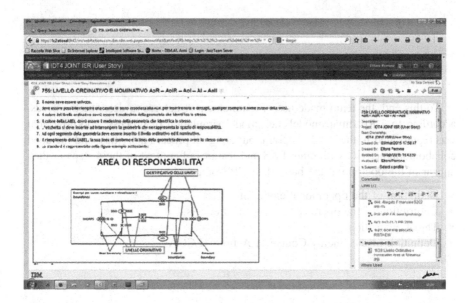

That led to the need of adopting requirement classification typologies that do not belong to the dominion of the *agile* methodology, to support the team in the following:

- Focusing on the FAS objectives;
- Providing a background context for the topics of each single *user story*;
- Defining the application environment and the objective of each phase, for a reuse in a further phase when the outcomes are achieved and information are passed on the JISR client FAS;
- Cooperating within the various armed forces and services to standardize the operational process of IPB preparation providing all the details, as required.

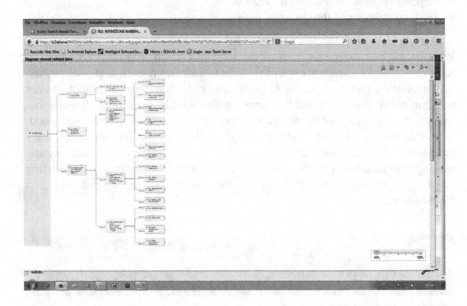

To this end, the team made use of requirement structures to describe the needs, according to a Roadmap model, i.e., model simulation of process workflow, that was represented in the block diagram, outlining all process activities, and included collaboration activity with other FAS and features of context of the 4 steps as they appear in the doctrine at the base of the IPB preparation:

- Definition of the operational environment,
- Description of the operative environment effects,
- Threat evaluation,
- Definition of the Enemy Course of Actions (ECOAs).

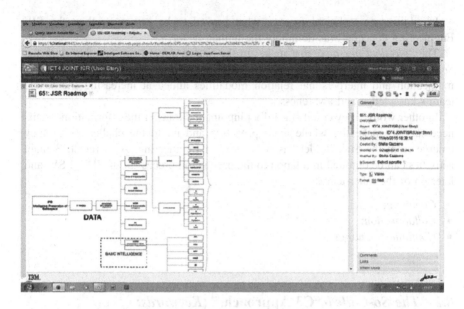

Use cases—this technique was used for the extensive and clear description of the operational flow in relation to a particular activity concerning the IPB preparation. This allows to specify the details of the operational scenario, as well as to clarify how the various players interact to support the chain of production. This chain stems from the Joint ISR activities and develops through processes and field activities of the privileged *intelligence* users, i.e., artillery and Military Engineers and CIED.

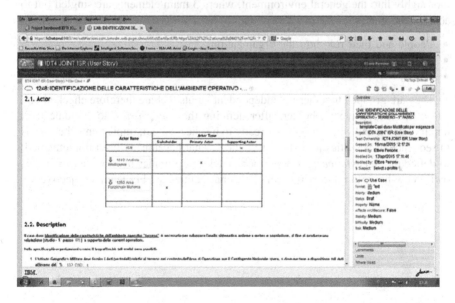

In addition to technical factors, what proved to be key more than the SW Engineering concepts was definitely the human factor.

The *"physical persons"* that make up the *Stakeholders* team together with the *Scrum Master* have painstakingly worked in cooperation to implement the communication and interpersonal relation modalities aimed at increasing the whole team's knowledge and awareness.

In other words, they could skillfully pinpoint and transfer the information details necessary to make the whole team proactive and up to the challenges as they emerge from the Stakeholders as well as from Engineering. As a result, 3 main activities can be outlined in relation to the production process of the IDT 4 SW and later on of the other teams:

- *Coaching*;
- *Collaboration*;
- *Continuous Change*.

8.2 The So-Called "C3 Approach," (Keywords: Collaboration, Coaching, Continuous Change)

When it comes to launching a project, due to the partial knowledge of what is going to be implemented, how and with what resources, some may have anxiety symptoms, and others may be excited at the idea of facing the challenge of a new start, depending on the personal inclinations. The winning approach consists in getting thoroughly into the general environment, where 3 main elements are singled out to achieve the goal:

- People and their role into the organization.
- Actions.
- Goals.

They are grouped together in "independent" units that are therefore checked out.

To this end, the Stakeholder, after defining the *user stories,* to evaluate their depth of detail and whether they are ready to be checked out, implements the *scripts* (a sequence of activities to carry out to check what has already been implemented) to define the *Acceptance Test.* Each personnel unit provides his own help with his own know-how, to build part of the item that defines the big team's objective.

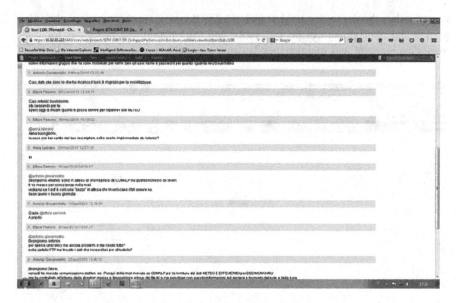

The functioning mechanism is user friendly, but the main factors at the core of a smart team are Coaching, Collaboration, and Continuous Change, as we already highlighted.

After each single loop, the *Stakeholders Team* evaluates whether the topic the *user story* is about needs to be introduced to the developers' team by a short informative lesson, laying the stress on specific details if necessary.

Each user and each team cooperates to achieve the goal; thus, each one provides his own piece of information that helps the team pursue the final objectives.

The tools for the JAZZ *Collaborative Application Lifecycle Management* allow to reduce the communication problems, which are typical of team geographically widely scattered. As a result, these tools are of big help to restrain conflict levels.

In particular, thanks to the IBM RATL JAZZ, the *task* synchronization, the integration and management of people, tools, and information and processes for *software* development and maintenance became a possible perspective.

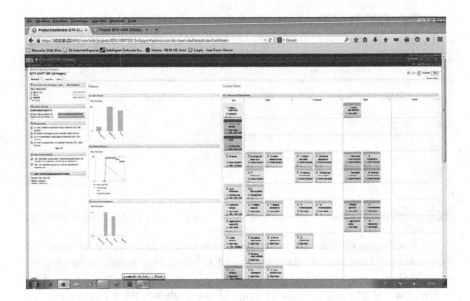

The Continuous Change process ensures the correct approach to the topic complexity evolution, allowing the periodic *tools* "refresh," as well as a reconsideration of the techniques and rules to overcome setbacks that may come across throughout the implementation.

Command, Control, and Communication are the 3 hinges of the collaborative approach to the art of military command, and they are supported by the IBM RATL JAZZ platform tools.

To sum up, the IBM *Rational* can effectively support the Agile through the following:

- A pragmatic approach for *Agile practice* meant to the incremental product measurement and the enhancement;
- An integrated technology platform that fosters the following:

  - Collaboration regardless of the geographic location, to facilitate an effective communication;
  - Automation of the maximum amount of activities to cut costs and reduce the likelihood of human errors;
  - *Reporting*, not intrusively drawn up in relation to roles for a general monitoring of the projects trend.

- In this collaborative development environment, marked by the interaction of the various *scrum teams*, the availability of a tools platform appears to be essential for a fast information sharing on the web.

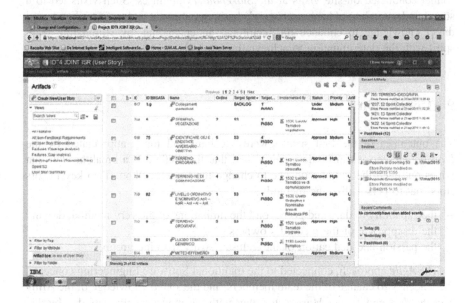

In this context, the JAZZ platform proved to be flexible and easy to hand out thanks to its tools. It stands out as an useful help to manage both the C3 tool (Command, Control and Communication) and the traditional activities of the collaborative development, which is the *Italian Army Agile*, i.e., the *agile* doctrine as it was made suitable for the peculiarity military environment.

As a matter of fact, the platform philosophy consists in automating the reiterative operations which are typical of the agile methodology. Such operations can be unmanned or they may tend to increase as the team members and teams increase.

In particular, the platform proved to be useful to the following:

- Make the *teams* and *Stakeholders* interact to capture and draw up the *user stories*. Actually, for hierarchical and organizational reasons, the *Stakeholders* are located in geographical areas other than Rome; as a result, this would become detrimental for the correct and timely identification of the *user stories* which will be assigned to the developers afterward;
- to monitor possible gaps between the planned workloads in terms of assigned *tasks* received by each *team* member;
- identify overlapping working areas of different *teams* for the effort of maximization and capitalization;
- share and specify the information elements;
- outline the system production cycle;
- carry out an impact analysis of a possible change with derivative cost evaluation.

In addition to a "conventional" use of the platform, civilian and military personnel conducted on-site visits at the *Stakeholders* premises. Such visits led to a sound understanding of each *user story's* basic requirement and to foster bonds and direct knowledge to enhance the reciprocal understanding, by stimulating team spirit with the determination to achieve the set goals.

9 Conclusion

The outcomes of the first sprints were excellent and much appreciated by the *Stakeholders,* so it can be definitely stated that the method is a breakthrough that wipes out even the initial reticence to change for a new method. Two key factors stood out to personalize the method:

- the capability to dare and go beyond the limitations of the very method, defying the disapproval of the method's owners;
- the passion and dedication of the industry/Army working group.

The *Agile* approach, connected to the traditional Software Engineering concepts, the tendency to change of the team members, and the sound knowledge of the Army *Stakeholders* and *Scrum Masters* together with the industry engineers day by day allowed to improve the *translation of the doctrine into Engineering*.

On the one hand, this means that the method works, and on the other hand, we cannot but underline our regret toward some unpleasant disagreements and clashes among the team members, originating from occasional conflicts of hierarchical, organizational, cultural, and sometimes "*political*" nature. This is the effect of a reluctant state of mind.

We had the feeling that such reticent attitude toward change, unless it is adequately managed in a diplomatic way (as it happened), may be of detriment to the whole project and cause a total denial with a relevant *rejection reaction* on the basis of misconceptions and cognitive *biases*.

Therefore, the lesson learned suggests that to prevent what has been described above, the method needs to be presented outside the team, comprehensively, thoroughly, and consistently, with pros and cons mirroring the actual achievements, so as to appear an appealing and credible method, as well as a desirable one if possible. On the other hand, it will need special care and a pragmatic approach for the daily activities, which builds upon the life experience, professionalism, and the needs of the occasional interlocutor.

As Niccolò Machiavelli said in 1513:

> And it ought to be remembered that there is nothing more difficult to take in hand, more perilous to conduct, or more uncertain in its success, than to take the lead in the introduction of a new order of things. Because the innovator has for enemies all those who have done well under the old conditions, and lukewarm defenders in those who may do well under the new. This coolness arises partly from fear of the opponents, who have the laws on their side, and partly from the incredulity of men, who do not readily believe in new things until they have had a long experience of them [6].

To sum up, as we anticipated in the abstract and in the introduction, the team-centric nature of this method as well as the peculiar military tendency to focus on group dynamics ensured to build a close-knit team, willing to tackle the challenges that will be arising throughout the future activities.

References

1. Schwaber K (2004) Agile project management with SCRUM. Microsoft Press. ISBN 9780735619937
2. Messina A, Cotugno F (2014) Adapting SCRUM to the Italian army: methods and (open) tools. In: The 10th international conference on open source systems San Jose, Costa Rica, 6–9 May 2014. ISBN 978-3-642-55127-7
3. http://www.jazz.net
4. http://www.nato.int [STANAG 4607]
5. Kreitmair T, Ross J (2004) Coalition aerial surveillance and reconnaissance simulation exercise 2003. In: Command and control research and technology symposium (CCRTS). U.S. Department of Defense (DoD), Command and Control Research Program (CCRP), June 16, 2004
6. Niccolò MACHIAVELLI "*Il Principe*", Chapter VI, a by Giorgio Inglese and Federico Chabod, Einaudi 2006. ISBN 88-06-17742-7

Role of the Design Authority in Large Scrum of Scrum Multi-team-based Programs

Giovanni Arseni

Abstract This paper discusses the vital role of the Design Authority (DA) within large Military Scrum of Scrum multi-team-based programs, in assuring that systems would support a military operation according to organizational and administrative procedures. It is a matter of fact that multiple teams and progressive technology can create a thwarting environment leading to implementation fatigue; finding balance between agility versus standardization or functionality versus intuitiveness often leads to discussions among the stakeholders. The DA manages these problems and maximizes opportunities, taking the responsibility for ensuring a solution that meets goals, needs, and specifications during the entire Agile process. A discrete DA will act in coordination with the teams sprint-by-sprint, defining and communicating portfolio vision, technical strategies, architecture standards, and design methodologies. DA maintains an end-to-end view with a mission-focused perspective, ensuring the delivery of business value and providing timely and useful information to team members and stakeholders; absence of a DA yields uncertainty among project participants on their tasks and roles.

1 Introduction

Transformation programs are important for creating business value and building strategic capabilities across organizations. With many enterprises as well as military organizations spending around 50 % of their IT budget on application development in order to fulfill stakeholders' requirements, the ability to execute software

Giovanni Arseni: Italian Army General Staff.

G. Arseni (✉)
Italian Army, Rome, Italy
e-mail: giovanni.arseni@esercito.difesa.it

© Springer International Publishing Switzerland 2016   181
P. Ciancarini et al. (eds.), *Proceedings of 4th International Conference
in Software Engineering for Defence Applications*, Advances in Intelligent
Systems and Computing 422, DOI 10.1007/978-3-319-27896-4_15

programs faster and at lower cost is essential to success for many transformation projects. However, large software projects, on average, run over budget and over schedule, as reported by McKinsey and Oxford University joint study [1], and many of them go so badly that will never go into production, because the software developer abandons the project, the owner cancels the project, or the developer delivers a system that the owner evaluates inadequate and unacceptable. In many large organizations, including the Armed Forces, efforts go on for years, with a lot of money spent and little delivered.

Ultimately, the purpose of software development in the support of evolutionary process is to deliver high-quality, on-time, and on-budget software to satisfy expectation, furthermore allowing for continuous sensible enhancements and developments. A blend of agile focus on delivery, human-centric support for customers and developers, incorporating dynamic requirements, and avoiding over-documenting and over-engineering exercises seem to be of benefit to software practice and consequently to the stakeholders.

Large-scale application development projects are particularly challenging because of their complexity and high degree of interdependency among work streams and related work packages. These projects demand close coordination and synchronization due to frequent refinements starting from the original vision given by the Product Owner (PO) and the subsequent adjustment in a repetitive cycle of continuous refinement [2].

Such coordination and synchronization can only happen by breaking down the traditional vertical tower in iterative application development approach with user requirements divided and organized into a number of parallel work streams and packages considering technology, protocols, and operational domains.

It has been proved that building highly complicated system imposes the use of cross-functional teams to break down the silos. Iterative application development practices inspired by agile shall be considered in complex projects such as large organization transformation; however, these approaches need to be carefully oriented when managing cross-functional team.

Multiple development teams using imposed progressive technology can create a frustrating environment, leading to implementation fatigue. Finding balance between agility versus standardization or functionality versus usability may generate harsh discussions among the stakeholders. Decision making can stall when visibility and transparency are limited. Creating a design authority can help to avoid these problems and maximize opportunities.

2 ITA Army Agile

Agile software development methodology addresses above-mentioned complexity and therefore has emerged as the organization's choice for an innovative software approach. Traditional software development is founded on system fixed requirement specification documentation which describes and details the capabilities that the system shall deliver at the end of the contract. Standard phases for a software project are analysis, design, code, and test. Agile software development methodology still uses these phases, but considers them also as continuous activities and not as fixed stages. These activities are mapped into an agile cycle, more specifically a scrum process. Scrum is an agile declination with short and fixed length of the Sprint production cycle—time-boxed sequence—(about 3–4 weeks) and a clear definition of the roles. During each Sprint, the above phases are implemented by the team as an iterative process able to deliver result after each Sprint.

On a traditional software project, there is 100 % risk and 0 % business value delivered until final day of the project when software is released [3]; the Scrum approach provides, on the other side, by delivering quality functionality incrementally, timely business value to the customer in weeks rather than months, and the risk is reduced thanks to the faster feedback cycles. Using an iterative approach, requirements are adapted using feedbacks collected with sprint reviews after each sprint, reducing the implementation gap.

The Italian Army, since 2013, adopted Scrum Agile methodology for developing Core and Mission Critical applications, as defined in "Cloud Computing Automating the Virtualized Data Center" [4] Core applications establish a *competitive advantage*, and Mission Critical ones provide a *temporal contextualization*. These classifications might also support in determining whether an organizations want to consume a "black box" solution form a third-party vendor. At the same time, in-house applications that contain intellectual property of the enterprise, system of innovation, would require *characterization*. There is a lack of semantics to define and characterize an application workload; meanwhile, new kinds of application workloads are emerging with mobile computing (MBaaS) and big data applications, now adopted for a defense organizations. We evaluate this as an N × N problem, if there are N number of application workloads and N number of the types of software and hardware architecture to provision them on. Day by day, we discovered that the key to success lies in the fundamentals of understanding the business workload and provisioning accordingly. Within ITA Army, SCRUM Agile is used to develop the Enterprise Applications, Land Command and Control Evolution (LC2EVO) able to interface or integrate with other enterprise applications, or other Software COTS, GOTS, NOTS [5] to be deployed across a variety of networks Internet, intranet and coalition networks, FMN [6], while meeting strict requirements for security and administration management. Italian Army during LC2EVO implementation discovered that original Scrum Agile needs to be adapted in order to cope with above-mentioned *characterization* process.

The Italian Army adopted a customized methodology, "ITA Army Agile" (IAA), based on the Army organization complex structure using the following actors: the product owner team (POT), which is a multi-users team in charge to manage Product Backlog involving stakeholders and domain experts with a mission-focused perspective managed by the DA that maintains the end-to-end view; the Integrated Development Team (IDT), a developer team including both military and civilian personnel, where all members contribute to all the phases of the project (development, documentation, testing, ...), interacting day by day with the DA in order to implement technical strategies, architecture standards, and design methodologies; the IAA Master (Scrum Master), which is a military member coaching and supporting the IDT as methodology expert and pursuing Product Backlog's Sprint Segment implementation; the IAA Coach plans and manages required resources Sprint-by-Sprint according to established financial and contract constraints.

3 Software Architecture and Documentation

Large Enterprise Software Solution requires complex software and infrastructure horizontal architectures including many components from third-party systems incorporated into the new whole. Understanding and validating an incremental set of requirements may necessitate to continue reasoning and modeling with a large variety of architectural solutions. Some requirements may need to be traded off against others to make the overall system feasible, sustainable, interoperable, and compliance. It has been found in many situations to be much better to do this in advance of a large code base and complex architectural solution.

The Agile Manifesto [7] pursues the value of working software over comprehensive documentation, but it also explicitly states that documentation can be valuable as well. Through the life of the project we found that at the beginning, with less components and applications in place, documentation was less important; however, integrating software components and dealing with IT increasing infrastructure at certain point, the need for communication exceeded the limits of face-to-face communication—both spatially and temporally. When hundreds of developers and other stakeholders are involved in the project, some documentation practices must be established to distribute the information efficiently throughout the organization. Besides, if the life span of a software system is over some years, then documentation is needed to bridge the communication gap between different generations of architects and developers. On the other hand, there may even be compelling compliance and operational and legal reasons that dictate documentation. In particular, core and mission critical systems must often evaluate risk, potential threats, and vulnerabilities that might be exploited and pass certifications that are largely based on reviewing documents, following a classification taxonomy that will adhere to different standards as follows: for corporate enterprise ISO 27001 [8], for Italy national laws and rules based on common criteria and ITSEC standards [9], and for Allied Operations following NATO regulations [10].

The above-mentioned set of documentation should not replace face-to-face communication when the latter is more appropriate. All documentation (including architectural) is written for two reasons: to communicate and to remember solutions. However, documentation is one-directional communication, and it is not a very efficient way of communicating among people. The most efficient way of communicating is more people talking face to face at the whiteboard. The main motivation to produce architectural documentation is to record the design and its rationale as a kind of collective memory. In particular, during the life of our project, the proliferation of applications deployed on Web 2.0 model architecture imposed to follow several authentication procedures. IT architects evaluated that the more complicated the access to data and applications, the greater the risk of losing privacy and integrity [11]. Therefore, the solution was to adopt a unified process for access to several applications commonly known as single sign-on (SSO) and identity management (IdM).

Software Architecture Documentation can have three major challenges:

- Size. One of the main concerns is that the amount of required documentation is rather large and the architecture document soon becomes unreadable and non-maintainable. To solve the problem, the document should split into smaller chunks;
- Fragmentation. In general, fragmentation a lot of existing architectural information is stored in different places (e.g., presentation slides, e-mails, and meeting memos);
- Separated architecture documentation. Architectural information was produced throughout the development, but architecture documentation took place only at certain major milestones and sometimes not well documented This caused a delay in the recording of architectural information.

The conclusions drawn from the experience were the following:

- Architectural information should be organized in small information packages that address stakeholders' needs (i.e., Standard Operating Procedures, Instructions, Field manuals, training pamphlet, Setup), rather than in conventional all-purpose, comprehensive architecture documents. The consistency of such information packages should be preferably guaranteed by the underlying infrastructure rather than relying on manual updating;
- The storing of architectural information should be centralized in a common secure repository, in order to share and distribute documents and reports, improving operational efficiencies by streamlining communications, enabling documents to be published as Web sites, and maintaining accurate, timely, and current Web content;
- The codifying of architectural knowledge should be seamlessly integrated with the agile process so that architectural knowledge is largely codified as a side effect of those activities that create the knowledge.

4   Design Authority Concepts

4.1   The Design Authority

"Design Authority" is an often used term. It is an important idea that has implications for the management and execution and responsibilities for a product design. First DA activity leads to enhance compliance, interoperability, and integration activities following governance guidelines.

Italian Army is engaged in National or Coalition operations and other missions that involve the development, integration, or use of Owned or Common Software deployed over a National or Mission Network. These conditions will require the availability of harmonized and interoperable and, in some cases, certified software baseline. It requires centralized configuration control and synchronization between software tools in order to implement the new paradigm of providing software as a service on a common infrastructure, i.e., NCI Agency as the NATO Design Authority established a framework for NATO Community Development (NCoDe) [12].

DA will act as advisor and facilitator with multi-teams with attention to:

- deliveries;
- deadlines; and
- final product integration.

Standard Agile bibliography suggests to sync sprint releases belonging to different development teams. It is also suggested to put forward team with more complex requirements or additional added value.

DA meetings should be held considering release dates starting since third sprint. Normally, DA meetings are organized before each steering committee meeting to report relevant technical issues.

DA should produce following artifacts:

- Meeting reports;
- General compatibility plan.

General compatibility report should be used for complex project involving several development teams.

This plan should include the following:

- Hardware and software specs or best practices included in the project;
- Interoperability standard required;
- Access standard required; and
- Arguments and formats that will be adopted.

This DA role is ever expanding particularly as there is more and more rapid work or agile work being done in the software/Web development space.

Every organization needs to set up their DA in the way that best suits the purpose to fix many integration issues since project initial phases or during his life cycle in order to make decisions in time.

4.2   Definition

The objectives of DA are as follows:

- To achieve project quality objectives, to improve delivery efficiency, and to optimize total costs;

- To control the high-level design concept, to approve designs, and to select standards;
- To communicate a standard design process to ensure interoperability, quality control, and integration of vendor components; and
- To consider opportunities throughout the projects steps, providing timely and useful information during the design process.

A DA is a group of people that:

- knows concretely how the project must proceed (vision);
- has the power to settle meetings and to implement choices (leadership);
- has the authority to make binding decisions (mandate);
- has the professional knowledge and skill to make architectural choices (knowledge);
- to motivates participants to perform beyond their own expectations (charisma).

5   Being Agile While Still Be Compliant: The Role of Design Authority in IAA

The DA role was assigned to the ITA Army General Staff by the Land Administrative Directorate. The responsibility was explicitly stated in the general requirements applicable to the entirety of the performance to be rendered with the contract at the base of LC2EVO, the new Army Enterprise Software solution funded by the "FORZA NEC" program. The first idea was that the DA members would act as advisors and facilitators versus the IDTs and the POT members, mentoring, coaching, guiding, and collaborating with them to reach design decisions that are supportive of current, as well as near-term state of the system.

The primary aim of the above-mentioned governance is to establish a high-level set of guidelines to the complexity of the ITA Army operational process, and, more, shareholders' protection and economic efficiency. The operational counterpart of governance is *compliance,* defined as "the process of meeting the expectations of others" [13], which shall consider a raft of different regulations. DA should carefully consider the relationship among governance, compliance, and operation as shown in Fig. 1 [14]. To be compliant imposes an assessment of the legislation and regulations with which the organization must comply.

All this implies that there are system features that are defined and decided outside of the IDTs; at this point, DA plays a decisive role. When products evolve, a big part of the development should consider existing functionalities, i.e., SSO and IdM. The running project cannot be left without enough effort spent gathering all the mentioned needs, and specially not only in the hands of self-organized IDTs. In order to successfully address and verify all the needs, it was established that reviews must not be only focused on the requested functionality and simply taking into account whether the coding is good and neat. Therefore, a development/integration

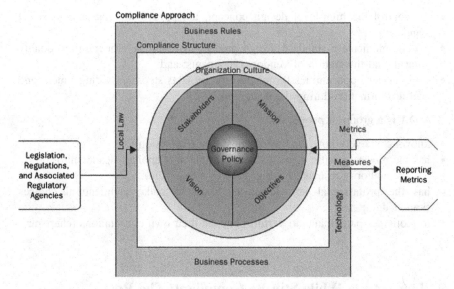

Fig. 1 Relationships among governance, compliance, and operation

environment has been built in order to understand whether the solution fits the entire environment or the architecture. Testing solutions have been extended over developers' unit testing. A large effort is made to participate in, or organize, large workshops or exhibitions, such as the SEDA, or multinational interoperability exercises [15] in order to verify the compliance to national and NATO standards and the capability to resist to cyber threats.

Managing the LC2EVO program, we found out, as a whole, that developers, which professionally ranged in a ranking from quite good to excellent, sometimes acted confined without understanding the complete picture. To make sure that the project teams could understand the big picture and the longer term vision, the DA effort is to make these clear, available, explained, and not only well documented, but also fitted to the architectural evolution of the whole. A big effort was done in order to have someone outside of the project that could review and accept/deny the solutions proposed: operational community and domain experts, production managers for maintenance, and management aspects.

One of last findings is that a DA in the Agile environment emerges naturally through greater perspective and experience rather than being asserted by hierarchy. It can only be effective with the consent of the teams. DA modeling identifies a similar role of the "architecture owner" who facilitates the creation of architecture rather than taking sole responsibility for it, defining a "hands-on" technical leadership role.

This model also offers an approach for scaling up development, where project architecture owners collaborate with the whole ITA ARMY CIS architecture efforts to ensure that appropriate strategies are being adopted. In this model, architects remain part of the team even at scale rather than forming a separate governance organization.

Design Authority will act throughout the Agile Development cycle bringing a considerable added value to the project, allowing budget and time frame control. While project complexity increases, it is strongly suggested to adopt an adequate DA.

In conclusion, we demonstrated that agile practices are not only compatible with existing severe governance of an organization, but it can be implemented in ways that provide strong support for compliance with given regulations.

References

1. http://www.mckinsey.com/insights/business_technology/achieving_success_in_large_complex_software_projects
2. Cotugno FR (2015) Managing increasing user needs complexity within the ITA Army Agile Framework. In: SEDA 2015, the 4th international conference in software engineering for defence applications
3. Goldstein I (2013) Scrum shortcuts without cutting corners: agile tactics, tools, & tips. Addison-Wesley Professional, Boston
4. Josyula V, Orr M, Page G (2011) Cloud computing automating the virtualized data center, Cisco Press, Indianapolis
5. Commercial Off-The-Shelf (COTS); Government Off-The-Shelf (GOTS); NATO Off-The-Shelf (NOTS)
6. FMN, Federated Mission Networking: The aim of the FMN Concept is to provide overarching guidance for establishing a federated Mission Network (MN) capability that enables effective information sharing among NATO, NATO Nations, and/or Non-NATO Entities participating in operations. A federated MN will be based on trust and willingness and will enable command and control (C2) in future NATO operations
7. http://agilemanifesto.org/
8. www.27001-online.com/, ISO 27001
9. http://www.ocsi.isticom.it/
10. http://www.infosec.nato.int/
11. "Network Security, Form risk analysis to protection strategies", Istituto Superiore delle Comunicazioni e delle Tecnologie dell'Informazione
12. NATO Communications and Information Agency (2013) Feasibility study on multinational NATO software tools
13. www.corporatecompliance.org
14. Conway SD, Conway ME (2009) Essentials of enterprise compliance. Wiley, Hoboken
15. i.e NATO Coalition Warrior Interoperability Exercise (CWIX). CWIX is a NAC endorsed, Military Committee directed and C3 Board guided Bi-SC annual programme designed to support the continuous improvement in interoperability for the Alliance. HQ SACT provides direction and management to the program, while NATO and Partner nations sponsor interoperability trials with specific objectives defined by ACT and National Leads. CWIX meets a broad spectrum of technical interoperability testing requirements

Make Your Enterprise Agile Transformation Initiative an Awesome Success

Enrico Mancin

Abstract Organizations that want to seize market opportunities, or simply prosper in a long-term time frame, will need to accelerate their innovation and delivery while reducing time to customer feedback. Agile transformations are complex and if you are leading an Enterprise Agile Transformation initiative, then you are surely running into challenges not directly addressed by Agile Manifesto or by advices provided by expensive luminary consultants probably not agreeing each other. Enterprise Agile Transformation requires the application of agile and flexible principles through the software development life cycle, as well as within the organization, to eliminate waste and delivery cycle time and unnecessary work, focusing on delivering value. Based on IBM real experience in successful large organization transitions, this paper will focus on the key aspects of enterprise agile transformation initiative and will help you on leading your transformation initiative at your organization in order to better prepare you to achieve a final success.

1 Introduction

Today, organizations around the globe are demanding for agility [1] to be responsive and fast in every interaction while expecting regular improvement to their experience. To stay competitive, to deliver great value with speed, simplicity, and continuous improvement, every successful organization needs to be agile. The major issue organizations have in front is that "agile" is not just a set of steps to be followed mechanically. Fortunately, today organizations can leverage existing and well-documented techniques as "way of doing" and can follow proven principles as a "way of thinking" in order to infuse new a mindset that can achieve true cultural change. Cultural change key challenge is that in "agile," we are never done because we are always asking "how can we get better?" Agile is not just a set of best

E. Mancin (✉)
IBM, Turin, Italy
e-mail: enrico.mancin@it.ibm.com

© Springer International Publishing Switzerland 2016   191
P. Ciancarini et al. (eds.), *Proceedings of 4th International Conference in Software Engineering for Defence Applications*, Advances in Intelligent Systems and Computing 422, DOI 10.1007/978-3-319-27896-4_16

practices, a term that implies there is only one right way to achieve outcomes. We are always changing to improve our performance.

Organizations are not agile are intended to lose in the battles with their competitors, gradually to lose the market war.

2 What Does Agile Look like for an Enterprise?

Agile is a new way of working. It was originally created to improve software development by shifting its focus to end-user requirements using rapid iterative development and prototyping. Based on the experience of agile, IBM professionals, working across the business, have expanded this traditional approach to encompass three core principles that apply to all professional fields. For simplicity, agile transformation initiatives should mainly focus on these principles:

1. **Begin with clarity about the outcome**, and let it guide every step along the way.
2. **Listen, iterate, learn, and course correct** rather than wait until it is perfect.
3. **Encourage self-direction for teams to unleash innovation**, instead of concentrating decision-making in the hands of a select few.

This "rule of thumb" is valid for every organization regardless its size.

Of course, "agile" provides us with the values, principles, practices, and traditions. Talking to agile teams, you will hear a variety of names for techniques used to apply agile principles. A short list of most significant "Agile Manifesto" values and agile practices is reported in Fig. 1.

At its heart, an agile team works in iterations: a series of short time-boxes of work (typically two weeks each). Everyone commits as a team to deliver certain outcomes by the end of each iteration.

However, I just want to briefly focus on some of the most common general practices adopted in "agile" organizations to apply "agile" principles, but it does not mean every team has to adopt them all or use them in the same manner.

1. **Begin with clarity about the outcome, and let it guides every step along the way**

   (a) **Story**: A basic unit of work that adds value to the outcome. Stories are written with a focus on user needs to be fulfilled by the work.
   (b) **Minimum Viable Product** (**MVP**): Defines a product by the core features required for the product to deliver a basic value to the end user. Over time, full feature development occurs using an iterative release cycle.
   (c) **Stand-up**: A short, daily meeting in which all team members announce what work they have done, what they will do until the next stand-up and any blockers (impediments) they face. These meetings focus on identifying impediments and ways to overcome them.

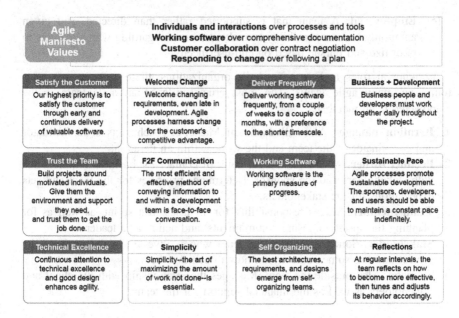

Agile Manifesto Values	Individuals and interactions over processes and tools Working software over comprehensive documentation Customer collaboration over contract negotiation Responding to change over following a plan		
Satisfy the Customer	**Welcome Change**	**Deliver Frequently**	**Business + Development**
Our highest priority is to satisfy the customer through early and continuous delivery of valuable software.	Welcome changing requirements, even late in development. Agile processes harness change for the customer's competitive advantage.	Deliver working software frequently, from a couple of weeks to a couple of months, with a preference to the shorter timescale.	Business people and developers must work together daily throughout the project.
Trust the Team	**F2F Communication**	**Working Software**	**Sustainable Pace**
Build projects around motivated individuals. Give them the environment and support they need, and trust them to get the job done.	The most efficient and effective method of conveying information to and within a development team is face-to-face conversation.	Working software is the primary measure of progress.	Agile processes promote sustainable development. The sponsors, developers, and users should be able to maintain a constant pace indefinitely.
Technical Excellence	**Simplicity**	**Self Organizing**	**Reflections**
Continuous attention to technical excellence and good design enhances agility.	Simplicity--the art of maximizing the amount of work not done--is essential.	The best architectures, requirements, and designs emerge from self-organizing teams.	At regular intervals, the team reflects on how to become more effective, then tunes and adjusts its behavior accordingly.

Fig. 1 Agile manifesto values and practices

2. Listen, iterate, learn, and course correct rather than wait until it is perfect

(a) **Feature presentation**: A demonstration held at the end of an iteration to display the outcomes of work done. Feedback is collected from the team, stakeholders, and potentially test users.

(b) **Retrospective**: A meeting typically held at the end of an iteration, in which team members celebrate what went well and speak honestly about work methods that could be improved in the next iteration.

(c) **Backlog**: A list of stories that a team wishes to develop in future iterations, in order of priority or dependency.

3. Encourage self-direction for teams to unleash innovation, instead of concentrating decision-making in the hands of a select few

(a) **Team composition**: Small, stable, cross-disciplinary, self-managing, empowered.

(b) **Co-located**: A team that works in the same physical space, to allow for immediate collaboration. Other teams use social mobility tools such as video conferencing and shared desktops to collaborate.

(c) **Cross-disciplinary**: A team composed of individuals with a variety of roles and skills. All are committed to working outside their given roles and skills to help teammates.

(d) **Self-managing**: A team that determines the work; it will perform in a sprint and comes up with its own strategy to meet those commitments.

(e) **Empowered**: A team that is supported rather than directed, and given autonomy and accountability to make decisions regarding which features to prioritize and pursue based on end-user needs.

Looking at roles in Enterprise Agile organizations, these are common roles found on many agile teams. The most crucial role of an agile team is the empowered team member:

1. **Iteration manager**. Also called a Scrum Master or coach. Acts as the facilitator and coordinator of the team and the work within an iteration.
2. **Product owner**. Acts as the single point of contact for business requirements with stakeholders and the sponsor. Determines priorities for the team and sets expectations for the stakeholders.
3. **Team member**. Shares responsibility for all tasks in a sprint, as well as for planning the sprint, choosing commitments, and acting as a leader.
4. **Stakeholder**. Someone outside the team who is invested in the goals of the project and has some authority on how to achieve them.
5. **Sponsor**. Also outside the team. Manages the relationship with the client and stakeholders as well as the financial support for the team.

3 Becoming an Agile Organization at Enterprise Level

What IBMers [2, 3] learned is that thinking, acting, and being agile is hard work. It takes courage, perseverance, and commitment. Starting by adopting a common language is crucial for a successful transformation and the language of design and development is the natural choice to be close to an agile mindset. Use words such as "design thinking" to describe how new client experiences and outcomes are conceived. Use words such as "systems thinking" to describe the art of making reliable inferences about behavior. Use words such as "experiments" to describe how you will challenge and disrupt the ways in which you work. What IBMers quickly realized is that this is more than a shift in terminology. It is a shift in thinking. And it represents fundamental changes in enterprises approach. This is an evolving and improving work.

When you lead an agile transformation initiative across a large organization you surely run into challenges not directly addressed in the Agile Manifesto or in well-documented agile practices reported in Fig. 2.

In addition, when you are looking for help, you have likely been exposed to inconsistent different suggestions and often expensive consultants who seem to agree on one piece of advice: "It depends."

Agile transformations are complex journeys, and without a clear understanding of what enterprise agile implies, your agile transformation initiative is at risk of

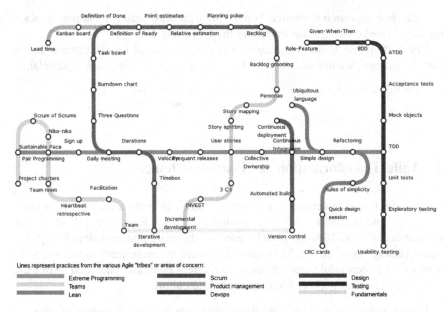

Fig. 2 The Agile Alliance practices subway map (http://guide.agilealliance.org/subway.html)

failure. This paper will deepen your understanding of enterprise agile and better prepare you to lead and participate in a successful transformation at your company or organization.

According to studies by Kotter, McKinsey, Blanchard, and IBM, what causes most organizational transformation failures are given as:

- Poor executive sponsorship
- Lack of employee involvement
- Ineffective change champion
- Underestimating the change effort
- Ad hoc approach change.

Let us look at what causes MOST organizational transformation failures.

Poor executive sponsorship. Often the transformation is not fully embraced at executive level.

Lack of employee involvement. Are we telling our teams that they are going to be agile or are we respecting that we have teams of motivated people and giving them a chance to self-organize and be a part of the transformation effort?

Ineffective change champion. We see that having change champions is critical. Liaison's who are influential and can help break down barriers.

Underestimating the change effort. An enterprise Agile transformation is more than implementing Scrum or whatever principle/practice. It is a huge effort and it is important to be aware of that.

An ad hoc *approach to change*. Not having a process, strategy, and plan in place for how to make change happen within the enterprise.

The bottom line is that the reasons most organizational transformations fail is that the changes are not anchored in the corporate culture. They are superficial.

> ...the failure to anchor changes firmly in the corporate culture [4].
> John Kotter

4   Agile Transformation Is "Leading Change"

Fortunately, this is something that has been studied in depth for a long, long time. An agile transformation is "leading change." John P. Kotter [4] of Kotter International is a well-known thought leader in the space and he defines 8 steps for leading change that (see Fig. 3), based on extensive research, needs to be completed in order for a successful transformation. These steps culminate and build so that the new approaches are anchored into the corporate culture.

> Leadership sustainability occurs when leaders accept why they need to improve, recognize what they need to improve and figure out how to make the improvements stick [5].
> Dave Ulrich and Norm Smallwood

Enterprise agile is really about company flexibility and agility in making your business. The broader and deeper agile is adopted throughout the organization; the more senior leadership must go beyond just "support" agility actively participating and leading. Nothing is more important that the matter of changes to middle level management where the effects of an agile transformation can be felt. In an agile organization, members that were traditionally informed through documentation and handoffs must now work in a collaborative way. Senior leaders embrace a new vision for collaborating across the enterprise and positively influence a new way to learn and improve. Senior leaders help to overcome obstacles to organizational change and they become a fundamental reference point for the others. An agile

Fig. 3 John P. Kotter "Leading Change" approach

1	Establish a sense of urgency
2	Create a powerful guiding coalition
3	Develop a vision and a strategy
4	Communicate the change vision
5	Empower a broad base of people to take action
6	Generate short-term wins
7	Consolidate gains and produce more change
8	Anchor the new approaches into the culture

transformation initiative starts from facilitating senior leaders and executives with quick workshops (effectiveness and optimization must be guaranteed for leveraging executives limited time) to create an exciting climate for developing business agility.

5 Agile Leadership Development in Practice

Let us see how IBM started to create leadership development experiences for IBMers who are on the path to becoming executives. Through the combination of in-person and virtual workshops, the team in charge quickly developed and deployed three new pilot offerings.

1. **Begin with clarity about the outcome, and let it guides every step along the way**
   The team started by creating an empathy map to get insight into the thoughts and feelings of their clients—IBM's aspiring executives. During this exercise, the team envisioned what an aspiring executive would think, say, feel, and do. This empathy map now serves as a compass for the team's work moving forward. Solution ideas that do not align with the vision are discarded; ideas that align with the map move forward.
   "Our vision keeps us moving toward solutions that help IBMers gain insight into who they are as leaders. When we default to approaches that don't align with the map, we discard them." Vicki Flaherty, Leadership Development (Fig. 4).
2. **Listen, iterate, learn, and course correct rather than wait until it is perfect**
   Initial ideas for creating a signature leader experience were posted to an Intranet blog. This provided members from around the world with the opportunity to share their perspectives and shape the team's thinking. During a design workshop, ideas were shaped into solution concepts. Of the six big ideas coming out of the laboratory, the team has developed three offerings:

   - a class for leaders who are nearing promotion to executive
   - a simple, easy-to-use job-aid tool called the Experience Maximizer
   - PALS "leadership circles," where small groups reinforce effective leadership practices.

Initial offering designs were tried, refined, and are now considered "living" offerings that will continually evolve based on feedback from clients.
"The design process invited us to "fail fast"—we did not try to create a perfect offering before we deployed. Instead, we came up with a solution that the team believed would address our users' needs." Jennifer Paylor, GPS People Engineer.

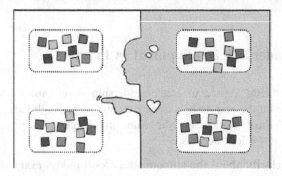

METHOD

EMPATHY MAP

WHY use an empathy map

Good design is grounded in a deep understanding of the person for whom you are designing. Designers have many techniques for developing this sort of empathy. An Empathy Map is one tool to help you synthesize your observations and draw out unexpected insights.

HOW to use an empathy map

UNPACK: Create a four quadrant layout on paper or a whiteboard. Populate the map by taking note of the following four traits of your user as you review your notes, audio, and video from your fieldwork:

SAY: What are some quotes and defining words your user said?
DO: What actions and behaviors did you notice?
THINK: What might your user be thinking? What does this tell you about his or her beliefs?
FEEL: What emotions might your subject be feeling?

Note that thoughts/beliefs and feelings/emotions cannot be observed directly. They must be _inferred_ by paying careful attention to various clues. Pay attention to body language, tone, and choice of words.

IDENTIFY NEEDS: "Needs" are human emotional or physical necessities. Needs help define your design challenge. Remember: Needs are verbs (activities and desires with which your user could use help), not nouns (solutions). Identify needs directly out of the user traits you noted, or from contradictions between two traits - such as a disconnect between what she says and what she does. Write down needs on the side of your Empathy Map.

IDENTIFY INSIGHTS: An "Insight" is a remarkable realization that you could leverage to better respond to a design challenge. Insights often grow from contradictions between two user attributes (either within a quadrant or from two different quadrants) or from asking yourself "Why?" when you notice strange behavior. Write down potential insights on the side of your Empathy Map. One way to identify the seeds of insights is to capture "tensions" and "contradictions" as you work.

= 15 =   d. ⊛⊛⊛⊛⊛

Fig. 4 Empathy map—Stanford University

3. **Encourage self-direction for teams to unleash innovation, instead of concentrating decision-making in the hands of a select few**
   For the design workshop, a small group from across the enterprise—most from outside of the Leadership Development organization—came together in a laboratory to develop offerings that directly address the needs of aspiring

executives. The entire team continues to work together, deciding what improvements should be included in the future iterations of these projects and programs.

"A strength of the agile design process is embracing collaboration and diversity of individuals in pursuit of a shared purpose. Also, it is incredibly empowering to be part of a larger team shaping ways that IBM's future leaders are developed." Elizabeth Silberg, Executive Assistant to James Kavanaugh, Senior VP, Transformation and Operations.

5.1 Why it Matters?

- To create differentiated client experiences—Bringing in varied perspectives early on and engaging a diverse group in an agile process increases the chances your products will resonate with your users.
- To grow as leaders through your own work—An agile process empowers teams to collectively make decisions.
- To restlessly reinvent your enterprise in a way that aligns with your corporate values and practices—To have new offerings to support aspiring executives and will continue to evolve them as you receive input from the users.

6 Building an Agile Culture

Continuous learning is at the core of an agile culture. Leading with agility starts with you [6]. It begins with making a commitment to your personal growth. Adopting an agile mindset and behaviors will engender the right outcomes. Only then will it be possible for you to lead and coach agile teams [7] and engage with clients in an agile manner strategy. The visibility is critical to maintaining the connection between planning and execution, and provides the information needed to make investment decisions and scope management.

Following are the rules for moving from a "command and control" mentality for achieving high performance by nurturing self-directed teams:

1. **It starts with you**
   Be a role model; live company agile Practices, be flexible, and listen.
2. **Help the team get more form themselves**
   Coach as coach-mentor, as facilitator, as teacher, as problem-solver, as conflict navigator, and as collaboration conductor.
3. **Get more for yourself**
   Recognize agile coach failure and success models, develop skills, and make your journey. Mastering feedback to engender trust and break down barriers to communication.

4. **Become agile using social techniques**

   You must restlessly reinvent the way you work, putting your client first, and unite to get things done, now. In this new way of working, social and agile techniques go hand in hand:

   - They enable people to work together, effectively and with self-direction
   - Both invite ongoing feedback and course correction
   - Agile starts with clarity of outcome, which is achieved by engaging the team openly and collaboratively

   Learn the social business techniques of transparency and collaboration that can help you embrace agility, especially among distributed teams.

7 Enterprise Agile Techniques

Enterprise agile techniques are typically performed when running experiments. Experimentation fuels agility. Clearly understanding stakeholders' needs opens the possibility for innovative solutions. You do not need to be focused on obtaining certainty and perfection, but you should experiment, iterate, and course correct tackling smaller pieces of the problem one by one.

7.1 What Is a Business Experiment?

- An experiment enables you to discover whether a new product or business program will succeed by subjecting it to a rigorous test.
- It can also give companies the confidence to overturn conventional wisdom and the faulty business intuition.
- The real payoff will happen when the organization as a whole shifts to a test-and-learn mind-set.

7.2 What Makes the Experiment Agile?

- In an increasingly digitized world, there is undoubtedly an increased need for speed and a need to align more closely with customers' needs.
- Applying agile frameworks, techniques and practices to the experiment, and eventually everyday working, will enable your organization to respond to these needs and achieve desired outcomes.
- Agile thinking is often rooted in software development with "sprints" (quick deliverables), continuous learning, radical collaboration, etc. but is increasingly being applied outside such settings

These are some techniques which can be used when running experiments; they will enable teams to work with agility:

1. **Framing Outcomes**—Helps teams clearly articulate what they intend to achieve during their experiments

   Defining specific and measurable outcomes, the team would like to achieve through the experiment. To be done at one time, at the start of experiment (can revisit and iterate through the experiment). Duration: 15 min overview on "what are outcomes" followed by 1–2 h to work on outcome statements and iterate on them until you reach a desirable outcome statement(s).

2. **Inhibitors Exercise**—Brainstorming technique to identify potential inhibitors and roadblocks that need to be addressed, solved for, or worked around during an experiment.

   Articulate the barriers and inhibitors that teams foresee in collaborating as one organization team and provide recommendations for overcoming the same.

   To be done at one time, at the start of experiment (can revisit and iterate through the experiment).

   Duration: 1–2 h in person/1–2 days through online jam

3. **Stand-ups**—Communication technique to get everyone in your team up to speed, to facilitate brainstorm discussions, and to enable collaboration. It can be seen as a more creative way of conducting status meetings.

   Helps in getting everyone in the team up to speed and enabling continuous alignment.

   Frequency dependent on engagement/ project context (daily, weekly, monthly, etc.).

   Duration: less than 30 min.

4. **Retrospectives**—Used at the end of any meeting or workshop to evaluate how things are going, capture feedback, and therefore help in continuous team/end-user alignment.

   Helps in capturing feedback at a point in time and therefore helps in continuous team/end-user alignment. It is especially helpful for coordinating multiple parallel activities in short time frames. It becomes the one source of truth or progress.

   Frequency dependent on engagement/project context (daily, weekly, monthly, etc.).

   Duration: 15–30 min.

5. **Decision Process**—Improve and hasten the decision-making process at your organization which typically is done at different speeds and varying degrees of effectiveness.

   Helps in improving the decision-making process by providing a clear and succinct format in making decisions and ensuring that the process is time-boxed.

   To be done in all decision-making situations

6. **Empathy Maps**—Used to better understand users/stakeholders which is key to help teams know-how to communicate and have the user in mind when preparing deliverables.

Collaborative tool teams can use to gain a deeper insight into their customers and stakeholders.

This can be done as needed, usually before preparing for a playback to the stakeholder/customer.

8 Conclusions

As already said above, active senior leadership is critical to transformation initiative success and organizational readiness is critical as well. The teams that do not work well individually do not work well together in collaboration. To allow teams to function well collaborating together, they must capable to perform well individually throughout foundational training post-training guidance and continuous coaching. So, the need for training and guidance applies to organizational leadership first. An organization has to cultivate the culture for agility throughout his leaders. An agile enterprise is not just about a group, a project, a program. It is not just a technical habit or attitude. It is all about agility of business.

References

1. The 9th Annual State of Agile Survey http://info.versionone.com/state-of-agile-development-survey-ninth.html VersionOne
2. Agile Development Governance. https://www-01.ibm.com/marketing/iwm/iwm/web/signup.do?source=swg-rtl-sd-wp&S_PKG=500005282&S_TACT=102GY5AW IBM
3. Agile Transformation—rethinking IT strategy in an uncertain world. https://www-304.ibm.com/easyaccess/fileserve?contentid=208473 IBM
4. Kotter JP Leading change. Kotter International
5. Ulrich D, Smallwood N (2013) Leadership sustainability. McGraw-Hill, New York
6. Quinn RE, Heynoski K, Thomas M, Spreitzer GM (2014) The best teacher in you. Berrett-Koehler, San Francisco
7. Coaching Agile Teams: A Companion for ScrumMasters, Agile Coaches, and Project Managers in Transition (Addison-Wesley Signature Series (Cohn)) Paperback—by Lyssa Adkins

DevOps Movement of Enterprise Agile Breakdown Silos, Create Collaboration, Increase Quality, and Application Speed

Francesco Colavita

Abstract The DevOps movement is gaining traction in organizations around the world because it is the best way to address many of the competitive challenges that enterprises are facing. It extends the most valuable aspects of agile development throughout the application life cycle, removing the bottlenecks that slow down application development and delivery, reducing enterprises costs, and improving satisfaction, results, and reputation. The best way to start this journey is based on an holistic approach that extends agile principles to the full enterprise; this model can help private and public companies to speed up the transformation and:

- Eliminate silos across the organization.
- Enable cross-team collaboration.
- Automate build, test, and deployment processes.
- Accelerate release cycles.
- Shift left in the development process and test in production.

1 DevOps Definition

DevOps is a movement in the IT community that uses agile/lean operations to add value by increasing collaboration between the development and the operations staff. This can take place at any time in the development life cycle when creating, operating, or updating an IT service. By using agile/lean techniques, it allows for IT services to be updated continuously so the business can capture market opportunities and reduce time for addressing customer needs.

F. Colavita (✉)
HP Italia, Via Ovidio 55, 00040 Pomezia, Italy
e-mail: francesco.colavita@hp.com

© Springer International Publishing Switzerland 2016 203
P. Ciancarini et al. (eds.), *Proceedings of 4th International Conference in Software Engineering for Defence Applications*, Advances in Intelligent Systems and Computing 422, DOI 10.1007/978-3-319-27896-4_17

Due to the age of the movement, it is not fully established and still prides itself on not being defined, but is really a set of practices that can be adopted by operational teams. This allows many people to serve many different roles in the field, but leads to confusion when trying to implement a DevOps culture in a traditional operational environment. In addition, the tools the community uses are still proliferating and have not been consolidated into traditional best practices. This consolidation will happen over time and the tools to support will become more robust/consolidated over time. The natural documentation of practices is starting, but have not been fully completed, so clear documentation of the community is not consolidated into standard practices.

2 A Brief History

In the mid-2000s, Enterprise Systems Management Movement came from an increasing need for innovation on the systems side of IT work. By 2008, there was enough momentum behind the movement to have the first Velocity conference occur, which allowed for best practice sharing regarding operations and for initial collaboration between developers and operations teams. Most initial collaboration was around providing updates to Web environments quickly and securely. In parallel to this, Agile Systems Administration was arising. The focus of this movement was on using lean manufacturing processes in IT systems. In 2009, the term DevOps was developed as a blending of these areas. The goal was greater collaboration between developers and IT operations to ensure as technology becomes more complex businesses can keep pace with what is needed for success.

As the DevOps community developed, it developed its own tools to help it accomplish its goals. Examples include Puppet, Chef, Juju, and Fabric (see DevOps for categorizations). The goal of these tools was to allow continuous delivery of systems and faster infrastructure optimization. The community was largely underground until it caught the notice of analysts at Gartner and Redmonk in 2011. This has led to larger interest in DevOps by enterprise companies. Gartner believes that by 2015, DevOps will be in 20 % of Enterprise business as part of the mainstream business strategy.

3 Who Is Adopting DevOps

As DevOps involves collaborating between various groups, DevOps workers can work in a number of areas. Sample titles illustrate this, including DevOps Engineer, DevOps Security Engineer, Application Delivery Manager, Automation Developer, and Tool Developer. They are often males under the age of 35 and often have an advanced degree.

3.1 Where Do They Work?

DevOps workers were originally thought of as only working in Web-based companies (e.g., Netflix & Google) and start-ups. Increasingly though large traditional enterprises are learning that they have to be more agile to address market needs and compete. It is difficult to incorporate DevOps into a system that has international financial, legal, and geographic considerations, but this is being done more all of the time. It is expected that 27 % of larger enterprises will have incorporated some level of DevOps by 2017.

3.2 How Do They Think?

Someone who works in DevOps is educated at a high level in both infrastructure operations and system development, but may have come from one of those fields and thus be an expert in this area. Based on the increasing complexity they have seen in systems, they have embraced a continuous improvement mentality. They are willing to learn new tools and have adopted tools that enable them to achieve this. They have a strong willingness to collaborate with developers and operations.

3.3 How Do They Like to Work?

DevOps people are tool savvy and use DevOps tools as a primary way to enable continuous delivery of system improvements. They use the data and information provided by these tools allows for data-based improvements for operation and development. In addition, they like bringing groups together to help achieve standard optimization of processes and systems.

3.4 What Motivates Them?

DevOps people are motivated by the ability to increase the value of IT services (internal and external facing) through:

- Driving to faster resolutions of business problems
- Driving increased customer satisfaction
- Enabling collaboration between teams to achieve a common goal
- Enabling multiple work streams to develop at once
- Using tools to monitor and improve performance
- Simplification of operational systems.

3.5 *What Worries Them?*

Worries vary based on the maturity of the IT environment. This can span the entire range of maturity. Here are two examples of concerns based on immature versus mature environments. When implementing a DevOps culture common concerns include:

- Breaking down the walls between developers and operations,
- Deploying code multiple times a day such as Facebook or Amazon without harming user experience,
- Working with operations to homogenize the data center hardware,
- Standardizing management tools,
- Working with developers to simplify the software architecture, and
- Minimizing deployment failures and the associated blame game that follows.

In a more modern IT environment where a DevOps culture is already in place common concerns include the following:

- Keeping communication flowing freely between developers and operations
- Increasing the numbers of developers on a project while ensuring accurate version control
- Automating tasks
- Identifying software infrastructure load issues to improve apps
- Increasing customer satisfaction.

4 Practices Adopted by Enterprise

4.1 *Developers Operations*

People who adopt DevOps principles work with both developers and operations to determine the needs of the business and use tools that simplify, synchronize, and automate current infrastructure and services housed on that infrastructure. Duties include the following:

- Continuous system improvement
- System release planning
- Continuous system integration
- Continuous update delivery

- Continuous system and infrastructure testing
- Continuous monitoring and feedback of systems and infrastructure for performance and security.

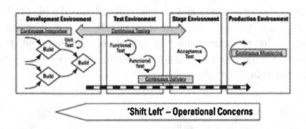

4.2 What Are the DevOps Myths

DevOps is a new name for something previously done in IT: There was not a large agile presence in the past and there were no tools to help simplify the development, implementation, and automation of apps. This arose as a need to address increasing need for speed to address customer needs and the complexity of IT systems.

DevOps gives developers the opportunity to do unlimited development: The key to this system is increasing the speed of development and the number of developers working on a project at once. The development should be based on customer needs and improvement suggestions gathered from infrastructure data monitoring.

Developers will understand infrastructure and operations will understand coding: This is not the goal of DevOps. DevOps is meant to increase the communication between the teams and make the overall process more agile.

Operations is not needed with DevOps; everything will move to the Cloud: DevOps does not have a goal to move primarily to the Cloud, but to a simpler, standardized infrastructure that can be more easily monitored for problems, deliver app updates more often, and be able to identify system optimization opportunities.

DevOps requires you to use certain tools: There are many DevOps tools and not all tools are used. The tools should be chosen based on the needs of the business. This will vary for every situation.

DevOps cannot be used in a large enterprise: Automation is a life cycle that has different needs based on the stage. It can be incorporated in an Enterprise regardless of the maturity of their IT system (see "What worries them").

We need to hire DevOps roles: DevOps is not a role; it is a way of doing things. A formal DevOps department is not required to implement a DevOps culture. You do need to adopt the culture to become more agile.

5   The Challenge: Keeping Efficiency in the New High Velocity

As we sow until now, to stay competitive there are the needs to accelerate the delivery of new software features and functionality; that is the idea behind the agile software development processes that are now widely used by application delivery teams to reduce delivery cycle times, but here is where things get harder. While agile methods are a huge step forward for application delivery teams, they alone do not ensure the fast rollout of new software.

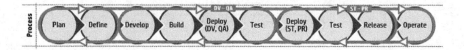

Cycle times are also driven by IT operations teams, which historically have been a speed bump in the road to putting applications into production. Why the delays? Operations teams want to avoid the risks that come with changes made to applications and infrastructure, such as breaching service-level agreements (SLAs), system downtime, and outages. So they approach new software releases cautiously. The result is as follows: New software releases are delayed and agility is lost at the end of the process chain. How to break up this bottleneck?

6   Driving Collaboration Across the Development and IT Operations

The lack of a standard definition for DevOps has created confusion for infrastructure and operations leaders who are trying to adopt this philosophy. There is no simplified approach regarding how an enterprise leader should start a DevOps journey, causing confusion about how and where to start. Each DevOps implementation is like a snowflake, what works for one may not work for another. By googling on the Web, we can find hundreds of definitions about DevOps and this does not facilitate companies that want to approach this new model. Let us see how it is possible to start in the right way this kind of project starting from one of the most shared definitions:

DevOps helps the organizations to bring together the key stakeholders (applications and operations) by focusing on collaboration, automation, and monitoring, resulting in improved application release velocity with quality.

So DevOps is all about people, process, and technology, and applying principles and methods to drive better collaboration between your software delivery and IT operations teams. DevOps extends the agile mindset to incorporate operations unit in it.

Whether your organization uses agile, waterfall, or both, DevOps recognizes the interdependences of application delivery and IT operations in accelerating the release of high-quality software. DevOps enables continuous delivery, which focuses on what is ultimately most important: shorter cycles for putting high-quality software into the hands of end users. Continuous delivery relies on better collaboration, integrated processes, and the comprehensive automation of build, test, and deployment processes to make software ready for release on demand.

7   An Holistic Approach to DevOps Implementation

While most DevOps initiatives are focused on leveraging automation and facilitate processes, it is critical to understand that the overall goal of a true DevOps implementation includes a tool chain; this concept enables a plug-and-play approach to evaluating and selecting tools, so that each tool can be loosely coupled to its adjacent tool in the application life cycle. Ensuring that all of the automation touch points are linked will result in speeding up the movement of releases through the tool chain while reducing errors, defects, rework, and outages between the information flows. Most modern integrations are being done at an API level but in some scenarios (e.g., passing models or policies), deeper integration may be required to ensure that specific parameters and data can be exchanged or shared cohesively.

Guidance: Tool chains are not for the faint of heart; a key element is the willingness to adopt a "stop the line" mentality, which comes from Toyota manufacturing, where any assembly line worker can pull a cord and stop the assembly line if something is wrong. The principle in development is the same: When there is a build failure, all work stops until it is fixed. Once established, the automated gates must be followed 100 % of the time. To the extent that loose integration is done, it is still critical to list all tools in your environment for each tool evaluation in order to understand if the vendor you are evaluating has an out-of-the-box integration or if you will be the one building it.

Don't stop there, however; also change what is not working. All process methodologies come with principles for continuous improvement, and DevOps is no different. Agile principles for continuous improvement leverage the lean concept of linking problems to approaches. Empowering DevOps staff to understand that outcomes drive behavior raises the importance of assigning meaningful responsibilities.

Many technology companies and service providers can help delivering pieces and parts of DevOps, and it is difficult to find who can offer the full solution that covers all the life cycle phases. To be effective and able to guarantee the success in global transformations, it is necessary to have, in addition to the technology, even expert people and process guidelines to drive the processes and implement them on the defined tools.

The experience in this type of projects, as for any project, is another essential ingredient that will drive the transformation in the right track. This means that any company that wants to start a DevOperations approach must deal with a deep market analysis in order to find the right provider that can guarantee all the mentioned aspects. It is very important to trust providers that can guarantee a complete software portfolio and a professional services organization, or good partners, skilled on the theme and the technology, able to help the customer to correctly implement the full solution.

8 Get on the Path to DevOps with Four Key Steps

ITIL continues to be the most popular IT operations framework available and is not bound by any specifics as to its use and application conversely DevOps challenges the conventional IT thinking. While ITIL adds more process rigor, DevOps seems to do just the opposite. It implies process, but allows each implementation to define and continually adjust to improve the desired outcome. It is this intentional vagueness that has stalled many IT organizations from implementing a DevOps strategy. While there is no specific set of required steps, we think it is important to define a cookbook to following in four steps that are the most critical to starting a DevOps initiative.

Many market analysts recommend to begin the DevOps journey by focusing on four keys to a successful transformation.

1. Assess the right DevOps strategy for you.
2. Identify the DevOps maturity of the core development and IT operations processes.
3. Determine standards and automation for continuous everything.
4. Establish measures and metrics for successful DevOps.

Step 1. Assess the DevOps strategy We have seen that DevOps emphasizes people and culture over tools and processes, and seeks to improve collaboration between operations and development teams. DevOps implementations utilize technology, especially automation tools that can leverage an increasingly programmable and dynamic infrastructure from a life cycle perspective. Having a common definition will ensure that those participating will focus on the project with a unified vision. Implementing DevOps will require significant changes affecting people, processes, and tools all in a tightly coordinated way, yet with a loosely coupled implementation that allows for a constant flow, enabling a constant feedback loop. Before beginning a DevOps initiative, an assessment of current systemic issues and technical debt is needed.

The As Is assessment and the To Be strategy definition help to define from the beginning what the company is trying to achieve and how much, in terms of cost and time, this transformation will require. In more specific terms, the assessment process helps the companies to:

- Identify the DevOps drivers, challenges, and initiatives
- Examine critical success factors and best practices from successful projects
- Review critical DevOps domains that align to the goals
- Identify priorities for short, midsize, and long-term projects

It is very important to include in the assessment the applications and services that will be involved in the transformation, while there are no specifics in the DevOps philosophy that negates applying DevOps to all types of applications, its core use case is in agile development, or Web-enabled applications. What makes these applications more viable for DevOps initiatives is that they are typically architected from the beginning to take advantage of all the DevOps definition facet lean, automation, programmable, and continuous delivery. Web-enabled applications are developed with lean and agile concepts, and often have nonfunctional requirements included, thereby improving manageability. Gartner and other analysts suggest categorizing the applications into systems of record in order to individuate easily the right applications to include in the new model.

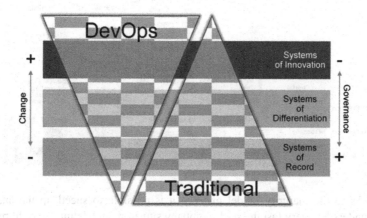

Step 2. Identify the maturity of your Development and Operations processes
To put the organizations on the path to DevOps, they need a clear understanding of the current practices for development and operations. Do the development processes use agile or waterfall methodologies, or both? Do the operations teams follow ITIL processes? What are the current practices for planning and communication?

In some specific areas of focus, the exploration of development practices looks at the approaches to application decoupling, continuous integration and the use of an applications hub, test automation, good coding practices, configurability, and maintainability. The review of operations practices explores the approaches to an operations hub, infrastructure as code, automated release management, resolution practices, and monitoring.

Ultimately, this process helps the organizations to:

- Identify key stakeholders in development, QA, and operations;
- Determine the used application development methodologies, agile, or waterfall;
- Determine the most important ITIL processes;
- Understand the planning and communication practices.

To understand the DevOps maturity of the core development and IT operations processes, the lead companies leverages a proven DevOps maturity model. This model looks at DevOps from three viewpoints—process, automation, and collaboration—and spans a series of clearly defined states on the path to an optimized DevOps environment.

The DevOps maturity exploration leaves the companies with an understanding of the organization's maturity level in terms of process standardization, automation tools, and collaboration approaches, along with insights into the opportunities for improvement.

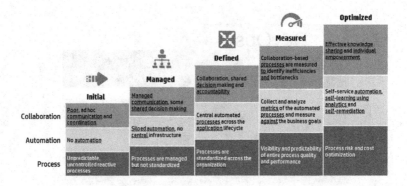

Thanks to the maturity model material it is possible to speed up the analysis phase, understand very fast the As Is company situation, and define the right proved steps to reach the To Be maturity level. A proven maturity model reduces the possibility to repeat errors that other did, because it collects all the experiences matured by the provider in other projects. It is essential to understand that any maturity model fits all recustomer requirements, but it is the best base on which to found the project personalization.

Step 3. Determine standards and automation for continuous everything
DevOps encompasses the entire application life cycle to ensure that applications can be deployed into production in an agile, automated, and continuous manner.

To that end, the DevOps journey determines the areas of standards and automation and then implements them in a phased approach. The ultimate goal was to move the organization toward "continuous everything," including continuous assessment, continuous integration, continuous testing, and continuous delivery. Automation is a key enabler of each these "continuous" processes.

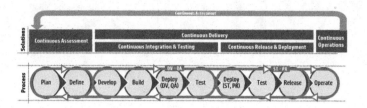

- **Continuous assessment** is used in the planning and definition phases of DevOps to automate the analysis of the impact of planned changes and to rationalize potential investments.
- **Continuous integration** and testing focuses on the triad of development, build, and test. This includes self-provisioning of development and test environments, automated build, and automated build verification tests. In addition, this "continuous" focus uses automated testing tools to help ensure that quality assurance work takes place throughout the application life cycle, including your agile development, staging, and production processes.
- **Continuous release and deployment** automate infrastructure provisioning. As code is compiled and built, infrastructure is automatically deployed and the configuration management database is automatically updated.
- **Continuous delivery** adds the ability to prepare software for agile releases—specifically, code that is ready to be deployed into production in an automated manner. Continuous delivery gives to the company the ability to release on demand—it allows releases to be driven by the needs.

Step 4. Establish measures and metrics for successful DevOps Another important key to the journey to DevOps is to identify the measures and metrics the Organizations will use to gauge the success of the initiative, such as needs enablement, cycle times, and escaped defects. In many cases, the DevOps metrics will be different from those used today in apps delivery and IT operations. The goal was to establish a common set of driven metrics that incorporates development, QA, and operational views.

Gartner offers this guidance: "Build new shared metrics that align with the needs and impacts, but, more importantly, help people realize that they must work together. Goals should be service-focused, with an eye toward improving agility (velocity) and improving value (quality)."

Bibliography

1. Gartner: Seven Steps to Start Your DevOps Initiative
2. Forrester: Four Steps To Win In The Mobile Moment
3. HP: Four keys to your Dev Ops journey

MBDA Extendible C2 Weapon System in Collaboration Environment

C. Di Biagio, P. Piccirilli, F. Batino, S. Capoleoni, F. Giammarino,
M. Ronchi, D. Vitelli and A.E. Guida

Abstract The need to improve the timescale and cost performance of C2 weapon system has led MBDA to identify a new way to deliver C2 weapon system based on a product line approach. An international MBDA team has been set up to identify commonalities in any command and control application within the air defence scope. This work led to an open architecture that encapsulates elements that depend on weapon system specificities by the ones that are common. The common elements, covering surveillance, threat evaluation, weapon assignment and engagement control functions within the wide MBDA's set of effectors, have been developed using MBDA Internal Research Funding and now are used to build up next-generation C2 solutions. The extendible concept behind the framework architecture allows the use of the 'MBDA Extendible C2' during the initial phases

C. Di Biagio (✉) · P. Piccirilli · F. Batino · S. Capoleoni · F. Giammarino · M. Ronchi
D. Vitelli · A.E. Guida
MBDA Italia SpA, Rome, Italy
e-mail: christian.di-biagio@mbda.it

P. Piccirilli
e-mail: pietro.piccirilli@mbda.it

F. Batino
e-mail: fabrizio.batino@mbda.it

S. Capoleoni
e-mail: sergio.capoleoni@mbda.it

F. Giammarino
e-mail: francesco.giammarino@mbda.it

M. Ronchi
e-mail: marco.ronchi@mbda.it

D. Vitelli
e-mail: daniele.vitelli@mbda.it

A.E. Guida
e-mail: anna-enza.guida@mbda.it

© Springer International Publishing Switzerland 2016   215
P. Ciancarini et al. (eds.), *Proceedings of 4th International Conference
in Software Engineering for Defence Applications*, Advances in Intelligent
Systems and Computing 422, DOI 10.1007/978-3-319-27896-4_18

of the contract (assessment phase, derisking, demonstrator, research) and for the deployment phase by providing the possibility for closed-loop feedback within the customer. The need to work collaboratively across nations leads to the design and delivery of the MBDA Collaborative Environment (or CEM) solution that is a collaborative engineering virtual laboratory for members who are located in separate sites. This environment is protected, which therefore enables the efficient sharing of data between Italy and other entities. This environment is a desktop-virtualization-based technology that creates a cohesive and scalable solution and centralizes the Italian infrastructure. This reduces development time as it allows real-time development and review of design artefacts, reducing the risk of duplication of effort. It reduces development risk of misunderstanding due to parallel development activities. Furthermore, it reduces time and costs of information technology set-up and management, and shipment time and cost. The CEM will be used to experiment Agile-distributed team collaboration.

1 Introduction

In the context of air defence systems, the command and control unit has the purpose of interfacing with human operators, external systems, sensors and effectors to accomplish the operational mission. The possible variations of those elements, for example the various types of radars interfaced by a specific AD system, have led to a large number of bespoke systems which show little communality, despite sharing a significant part of the underlying functionalities. MBDA has drawn on a long experience in producing weapon systems and instantiated a product line approach to this problem, making a rapid development feasible for future C2s.

Additionally, the nature of MBDA as a European company that is distributed on different sites, across United Kingdom, France and Italy but also Spain and Germany, with the need to reduce the delivery time, pushed the delivery of a protected collaborative laboratory: the MBDA Collaborative Environment, formerly known as CEM. The new development environment is really innovative for defence company, and it provides the following capabilities:

- It is shared and accessible from separate system
- It is able to support system and software complex engineering tools
- It is secured to exchange data
- It supports the exchange of any material using intangible export licence.

The need to work in a shared segregated environment comes from the experience that, for many programmes, there are few data more than restricted to manage. A good solution is to work with system models in a shared way until it becomes necessary to complete them with higher classified data: from this phase on, we will proceed inside a Tempest-compliant laboratory centre that allows the management of classified data above 'restricted'.

Moreover, in Agile development, the issue to be colocated could be a real barrier for the Agile adoption. The CEM is the solution to this issue.

2 Why Do We Need to Innovate?

MBDA's product portfolio has started to evolve from a bespoke development tightly tailored on specific customer needs towards a greater value for money and reduce time-to-market.

Classical design approach (see [1]) and function-oriented description (see [2]) fit perfectly to bespoke systems, but result ineffective in supporting reduction in cost and time-on-market, which can be obtained from the reuse of design elements already available in previous or ongoing programmes.

Analysis of several previous missile systems produced by MBDA has shown that air defence missile systems share a large set of functional requirements, yet differ along two main axes of variations:

The first is the interaction with the human operators, both in the details—i.e. what commands a specific operator does give, which information is exposed to him/her —and in philosophy, meaning the amount of authority the human operator has and their role in the weapon system, which can range from the fully automatic to the 'always a man in the loop' philosophy.

The other is the peer systems that are interfaced by the C2, both in their specific type (several types of sensors from different manufacturers; or conversely different types of missiles) and in their characteristics of usage in the weapon systems (how many sensors and which capabilities they do have, whether they are slaved to the C2 or are independent providers of information).

3 A Component-Based Product Line Approach

3.1 Goals and Approach of the Product Line

The goals of the product line approach are as follows:

- Provide a standard set of use cases that support an ordered set of user stories, which can be specialized into a given instance of the product line or extended.
- Provide a standard architecture that encapsulates the specificities needed to interface different real equipment, minimizing the impact of changes forced by the change of equipment.

- Clearly separate any presentation concerns, which are very dependent on customer's concept of operations, from functional aspects.
- Allow rapid prototyping of the solution. This enables a deliver-early, deliver-often approach and rapid loops with the customer to reach acceptable completeness on a certain set of use cases.

The approach followed to reach the goal is based on the identification of a set of components, divided in two main domains: the HMI domain and the server-side domain.

Within each domain, a set of components have been identified to provide contained, autonomous sets of services that, properly interconnected, can provide the required functionalities.

The middleware chosen for the implementation is an open standard, DDS (see [3]), that greatly facilitated integration of components.

3.2 Server-Side Components

The server side has, with regard to the initial goals, the onerous task of minimizing the variations caused by changes in the external equipment. This is a high challenge, as the coupling, e.g. between a sensor and a C2, is usually very strong, as they evolve at the same time, with the same functionality that sometime is implemented in the C2 and sometime in the sensor, and there is an implicit assumption that the C2 will use every function of the sensor. Even higher coupling is to be expected with regard to effectors.

3.2.1 Preliminary Analysis

A long analysis, taking into account several ground- or ship-based air defence systems, has been accomplished, producing a use case definition spanning the whole life cycle of the C2.

From that, it has become evident that, despite the specific features and technology of a given sensor, the interaction between the C2 and the sensor has the only purpose to provide the best possible tactical picture (and the notion of best can actually be quantified, based on characteristics of the tracks generated), with a special consideration for tracks under engagement (Fig. 1).

Similarly, the interaction with external links, despite the different message sets and interaction rules, is grounded firmly in the operational hierarchy of the C2, and different types of networks share the same set of fundamental purposes, that is briefly to allow surveillance coordination and fire distribution over a wide set of connected systems.

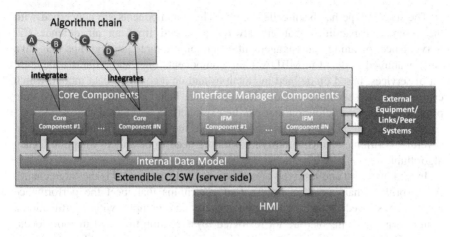

Fig. 1 Components in the MBDA Extendible C2 (server side)

A particular case has to be made for the effectors, as the obvious need to extract the whole possible effectiveness from the missile defence system means that the missile is accurately known and modelled inside the C2, and missile characteristic have a deep influence on planning, control and real-time modelling during flight.

3.2.2 Component Taxonomy

Given the result of the analysis, several types of components are needed to achieve the goals.

The first type is what has been called 'interface managers'. The purpose of an interface manager component is to interface a real equipment (of a certain type: surveillance sensors being different from IFF sensors, for example) and abstract the specific interface into a common, reusable set of services offered by the connected equipment.

It is notable that interface managers (IMs for short) are rather capable, perhaps more than what the name might suggest: aside from performing the task of message translation, from the equipment specific to the common data model, they also take care of performing those functions (initialization, synchronization, balancing) that are needed by the equipment yet do not contribute to the use cases of the system, going as far as automatically selecting some parameters based on the information published by other nodes. As a rule of thumb, we expect to have an interface manager in a C2 for each real external equipment that is connected to the C2.

Interestingly, when seen from the point of view of the server, the IHM is 'just an IFM', in the sense that a very limited set of assumptions is made on that component despite the large swath of functionalities implemented. This allows decoupling between the logic + algorithms and the presentation.

The second type has been called 'core'. These components are concerned with the essential capabilities that are always requested from an air defence C2. Surveillance, planning and engagement execution are three large groups that have been examined in detail in MBDA. Core components offer and request a standard set of services, based on defined interactions and on a common data model, which captures the rich semantics needed for a complex behaviour and at the same time provides a standard definition that can be exploited by all other components.

At the same time, the core components have the possibility to integrate one of several algorithmic chains, complying with the need to be able to integrate legacy algorithmic.

Indeed, the third and last type of components to be integrated is the 'algo chain'. An algorithm chain is a coherent set of algorithms that meet the performance requirements needed for the weapon system and comply with nonfunctional requirements in terms of (but not restricted to) execution time and memory occupation. Different algorithmic chains will provide different ranges of performance and sophistication of the input/output data and, even if they can be integrated into core components, should be used as a whole to guarantee a given performance level.

3.3 HMI Components in the Product Line

In complex systems, such as C2 weapon system, human–machine interface (HMI) systems provide human operators with the means to perform control functions and receive feedback on the undertaken actions. Design and methodological processes that should keep on board all technical, ergonomic and commercial application requirements have a great impact on both architecture and usability of HMIs soundness. Even though linking HMIs to the core application underneath can impact both operator and system performance, HMI design and development have been made independent of the implemented system and business logic—as seen in previous section on served side components—making them fit for reuse in different application scenarios. The technical framework architecture used is the result of applied Model-driven Architecture (MDA) (see [4, 5, 6]), where each capability represents an autonomous building block (ABB) and the actual integration of all components in a complex software solution is realized using the provided service interfaces.

In the exploded architecture, two decoupled but integrated macroblocks are, respectively, as follows: (i) Presentation Framework and (ii) HMI Foundation Framework.

Presentation Framework. It offers a widget library that can be integrated seamlessly into the presented architecture; in particular the presentation container component provides support for anchoring a third-party library (in the specific case, also said presentation framework). As it may be imagined, the integration process that interconnects the presentation framework with the presentation container component expects the development of a wrapper layer. Once such layer is developed, the presentation framework can be defined as completely integrated. The main responsibilities allocated to the presentation framework can be summarized as follows:

- It provides support for interacting with user, no matter which is the technical support (e.g. touch screen facilities provided to the underlying architecture) and/or the device (embedded or not).
- It provides support for maps (provided by a third party and out of scope with respect to the framework).
- It guarantees the ability to be independent from the device resolution and so on.

The listed responsibilities are inherited by the presentation container component that will have only the responsibility to translate the stimulus coming from the presentation framework towards the expected format by the underlying layers and/or components.

HMI Foundation Framework. Each instance provides several integrated capabilities (entirely managed by the nonmodifiable code of the software framework); for instance, logging and recording (i.e. the ability to store received command and reply them on-demand) management is example of services natively provided. Below there is a brief summary of all the other offered capabilities.

Data-centric Communication Support. The service layer offers publish/subscribe communication support, useful for several internal capabilities, for example the state replication among the deployed instances.

Multiple Instances Seamless Integration. Different instances of HFF may be integrated seamlessly using the data-centric communication support that is able to replicate the persistence layer: by means of reliable events exchange, the state of the instances may be aligned and when needed the integration may occur very easily; in fact, it is only needed to activate the replication mechanism and the instances will be integrated just sharing the state.

Multi-profile Support. An instance of HFF may be shared by several profiles. A profile is intended as a presentation profile, or rather as a different way of presenting the same state data: one instance with multiple presentation profiles, so such profiles share the same state and the framework takes care to implement the assured mechanism seamlessly for each profile.

This flexibility guarantees that HMI will meet common needs coming from several stakeholders and/or fit generic HMI requirements shared by several teams, groups and even companies, reducing product delivery costs and allowing to save budget (see [7]) (Fig. 2).

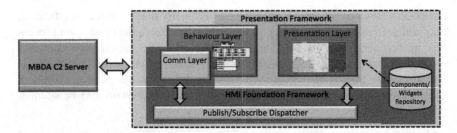

Fig. 2 Components in the MBDA Extendible C2

3.4 Usage of the Components

3.4.1 Case Studies

The product line has been used in three case studies as follows:

1. Tendering support: in an IT-UK programme, the PL has been used to develop a working demonstrator to assess the need directly with the customer and develop stories hands-on through interactive sessions.
2. International development: in the same IT-UK programme, the PL has provided the backbone of the development of two national variants of a Shorad system that differ for the physical allocation of the SW components, sensors and effector used by each. The demonstrator has been a first iteration (instead of a throw-away development). In this case, the PL provided the additional benefit of identifying the parts that are common for the two national customers.
3. Demonstrator for export contracts: as the product built from the product line can evolve, in terms of functionalities and HMI, incrementally and quickly, the PL has been used to build a demonstrator that is used in initial interactions with various export customers. The ability to quickly react to feature changes while keeping the architecture intact is particularly prized in this context.

For the first two case studies, the cost reduction stems from two improvements: the first being that a set of core components is already existing in the beginning of the project, and the second one being that the architecture allows to identify the components that are new (and therefore to be developed), yet are common to the two national configurations, and have to be developed only once. A saving in costs for the C2 of around 30 % of initial estimates is expected, with the two improvements contributing in equal measure to the goal.

In the third case study, the main advantage is that the turn-around time is very short, allowing tailoring—and tailoring for different customers—quickly, in order to secure possible contracts with a more solid technical understanding of the need.

3.4.2   Experimental Phases and Risk Reduction

The benefits of the MBDA Extensible C2 also extend to internal work. In premise, it offers a vocabulary that is unambiguous and agreed within all MBDA, along a complete SysML and UML model, with a set of use cases that capture the typical behaviour of a C2 working implementation. This allows a much quicker ramp-up of work in international teams.

This means that any experiment (e.g. comparison of middleware implementations; improvements in algorithmic chains; extension to interception domains that are not yet consolidated) will immediately find both a context and an area of work. The component structure allows a good partitioning of concerns; therefore, each improvement or change does impact not all the components, but rather a minor set; a working implementation can generally have every short cycle, obtaining a significant risk reduction when inclusion in target projects is deemed necessary.

4   The CEM Description

The CEM is a full redundant self-subsistent infrastructure based on 'VMware Horizon View' software that allows remote desktop virtualization and collaboration (Fig. 3).

The infrastructure is built up of one HP BladeSystem C3000 equipped with four HP ProLiant BL460 Blade servers, two HP Virtual Connect switches for external connections and two HP/Brocade Fiber Channel switches for connection to a dedicated HP P2000 G3 MSA Array Storage with 10 TB of disc space (Fig. 4).

The infrastructure is capable of providing all services needed to operate like a 'server room in a box' such as Active Directory for authentication, DNS and DHCP for networking, storage for file sharing, and server and desktop virtualization capabilities. The infrastructure is logically located in a DMZ of MBDA's network whose access is regulated through a firewall. Only one IP address and one port are exposed to MBDA network to provide services to final users.

Two VMware vCenter instances have been deployed: the first one is responsible for hosting and providing all infrastructure services to the CEM and the second one is responsible for hosting and providing all virtual desktops that are used by users to access data. First vCenter is reachable through the MBDA server management network, and access is restricted to IT staff. All infrastructure services, the second vCenter instance and all the virtual desktops are hosted in the dedicated DMZ network and reachable only through firewall access (Fig. 5).

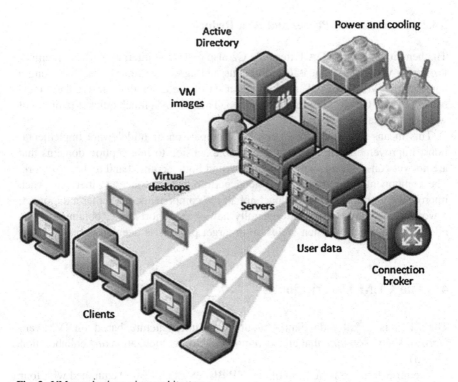

Fig. 3 VMware horizon view architecture

Italy and other users access CEM services through a client software called 'VMware Horizon Client'. The client, after logon, establishes communication with the CEM via Https-encrypted communication. The user is then provided with a virtual desktop that allows him to use the shared software development tool chain. Two virtual desktop tools have been created: the first one, based on a Windows 7 Italian image, and the second one, based on a Windows 7 English image. The virtual desktops, except for the language settings, are identical for Italy and other nation users. The virtual desktops are not persistent so, and after a user disconnection, the desktop itself is destroyed except for user data stored in the central file server. These settings assure security, consistency and manageability of the entire CEM solution (Fig. 6).

All systems and technologies are COTS such as to improve stability and supportability by external vendors and contractors.

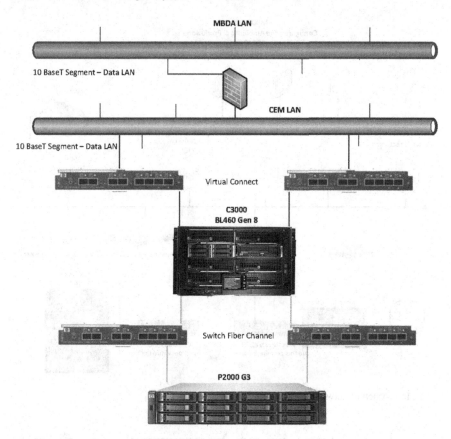

Fig. 4 Infrastructure physical interconnections

5 The Export Licence

According to the Italian law 185/90 on export of defence article, the exchange of technical data by intangible means with other Countries is subject to a specific licence.

Thus, one of the main objectives for MBDA Italy was to make the CEM compliant with the Italian Export Control Law requirements as follows:

- Ensure secure access with user ID and password
- Log each access
- Ensure the tracking of exchanged technical data and the related licence consumption
- Ensure reports availability
- Backup of all the contents exchanged

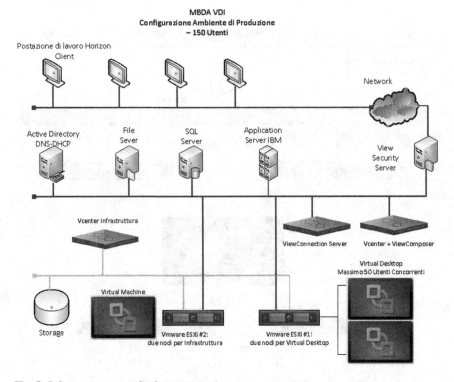

Fig. 5 Infrastructure network view

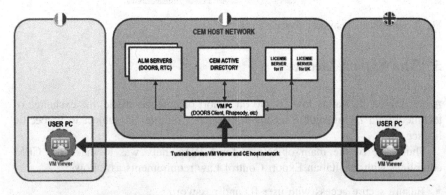

Fig. 6 Infrastructure logical view

For doing this, MBDA Italy put in place a specific internal procedure to comply with Italian regulations. It is one of the first instantiation in Italy of this kind of licence.

6 The CEM Security

The infrastructure guarantees that the CEM is separated from the commercial network by a firewall. The users, authorized by the MBDA security office, are provided for account/password to access the collaborative environment. The data are managed by a separate system.

All the procedures in order to ensure compliance with the legal obligations are described in the 'Regolamento Interno Sicurezza', and their application is guaranteed by the system administrator. The resources that have supported the development of the CEM has respected the need to know.

7 The CEM Benefits

The adopted solution brings benefits in terms of time as there is a single common master set of design and development artefacts. A single shared repository enhances coherency of design to meet all external and internal customer needs.

The CEM environment allows to share instantaneously information. With the legacy process, the time to share data is about ~5 days if all export processes are in place for each tangible item. The classified item must be burned in a CD-ROM and then carried to the MBDA IT security office. So it must be sent to the MBDA UK/FR security office which in turn will lead it into the computer centre enabled (Fig. 7).

With the collaborative environment, the process to share data is in real time (Fig. 8).

All steps in the management of updates, fixes, configuration, backup and set-up can be performed only once for all instances. Only one shared infrastructure optimizes software licences.

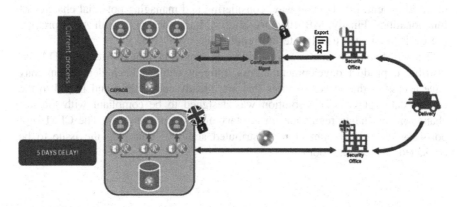

Fig. 7 Legacy process: time to share data = about ~5 days

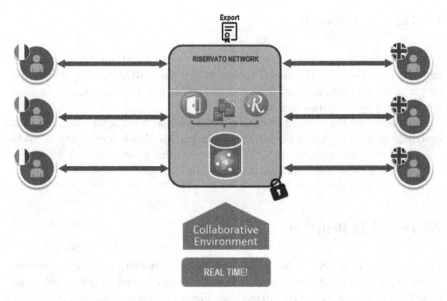

Fig. 8 Real-time data synchronization

8 Conclusions

The main objective of the MBDA Extendible C2 is to support a solution that responds appropriately to the needs of the customer. In the past, it was very difficult to define properly and completely all the requirements of a complex system such as C2 that the customer judged only at the end of the development. Starting with basic functionalities of C2 to extend, enriched according to customer demand, the solution has reduced development time, allowing to validate in advance functionalities and increasing customer satisfaction. In other words, this architecture meets the Agile approach, responding to the needs to build product from the early stages of development, in a flexible way, considering and managing potential changes of functionalities. Finally, MBSE/MDA- and UC-based design fit with Agile product life cycle versus the usual waterfall product life cycle [8].

The CEM solution creates an innovative environment, unique within MBDA, for distributed product development across different entities. Each department may have access to the necessary tools that are already cooperating and linked to the correct data sources. This solution was designed to be compliant with the last change in the Italian regulation on military goods export control. The CEM supports Agile [9] development for distributed activities removing the issue to be colocated in the same room.

References

1. DeMarco T (1979) Structured analysis and system specification. Prentice-Hall, Englewood Cliffs
2. AMDPCS UFD 1.0, U.S.Army ADA School, 1 Nov. 2001
3. OMG—Data Distribution Service for Real-time Systems Version 1.2—formal/07–01-01 http://www.omg.org/spec/DDS/
4. Bennetta J et al (2010) Aspect-oriented model-driven skeleton code generation: a graph-based transformation approach—designing high quality system/software architectures. Sci Comput Program 75(81):689–725
5. Stahl T, Völter M (2006) Model-driven software development—technology, engineering, management. Wiley, New York
6. OMG—Model Driven Architecture web site http://www.omg.org/mda/index.htm
7. Lazar I et al (2008) An Agile MDA approach for the development of service-oriented component-based applications. In: IEEE, 2008, first international conference on complexity and intelligence of the artificial and natural complex systems, medical applications of the complex systems, biomedical computing
8. Beck K et al (2001) Manifesto for Agile software development. Agile Alliance
9. Cohen D, Lindvall M, Costa P (2003) Agile software development. By Fraunhofer Center Maryland, DACS SOAR Report

Shu-ha-ri: How to Break the Rules and Still Be Agile

Dott Antonio Capobianco

Abstract Fata Informatica is a software house that produces a commercial system widely used by Defence Organizations. Our first approach to software development was standard waterfall, but during the years, our team adopted an agile-driven approach. The agile adoption was a long sweaty path that lead us through the Shu-ha-ri Aikido stages and let us implement our own agile framework. In this contribution will be shown as Fata Informatica's team defines the product increments that add the more value to the customers, how they develop and deliver them into the market. A description of our value and cost evaluation and the way we choose which increments add to a specific delivery, Fata Informatica's agile approach to the stakeholder engagement and how the static roadmap concept is outdated by a more agile roadmap definition approach will also be shown in order to meet the increasingly needs of our stakeholder.

1 Introduction

In Fata Informatica, it has been since 1994 that we deliver software. More than 10 years ago, we started a new software development line with the ambition to create a commercial product to sell in the IT marketplace. Almost immediately, we faced the problem of structuring a software development line able to produce patches and new releases of our product like all the most known softwares do.

We started our first release in 2002, and we adopted a waterfall design and development process. Even if we were able to build our product, we felt immediately that something was wrong.

I followed the PMI rules, and now I'm still PMP certified, but the result was a functioning product with extensive documentation written but the costs of the documentation production were comparable with the development costs!

D. Antonio Capobianco (✉)
Fata Informatica, Rome, Italy
e-mail: a.capobianco@fatainformatica.com

© Springer International Publishing Switzerland 2016
P. Ciancarini et al. (eds.), *Proceedings of 4th International Conference in Software Engineering for Defence Applications*, Advances in Intelligent Systems and Computing 422, DOI 10.1007/978-3-319-27896-4_19

So we tried to find out new solutions that could help us to be more cost-effective.

Our customers pays us for the software produced and not for the project documentation written!

We found articles on agile project management, and we fell in love with this "strange kind" of project management framework.

2 Agile Framework Adoption

We are now adopting an agile project management solution that is a mixture of Scrum, XP and Crystal.

We went through all the stages of the Shu-ha-ri process, the Aikido learning path, introduced in agile by Alistar Cockburn.

The Shu-ha-ri progression moves from obeying the rules (shu, which means to keep, protect, maintain), consciously moving away from the rules (ha, which means to detach or break free), and finally unconsciously finding an individual path (ri, which means to go beyond or transcend).

Initially, we adopted Scrum, and we followed strictly the Scrum rules. Our team followed several courses about Scrum, and we started to follow the rules by the book. We were in the shu stage.

After several months, we mastered the rules and we realized that the rules we were following were not so good for us. So we break the rules, and we created our agile project management model. We do not think that this is the best model, we do not know if others can successfully adhere to this model, but the only thing we know is that this models suits our expectations and our needs.

3 The Team

The basis of our development process is the team. An Agile team must be committed and self-empowered. Our initial approach to the project management was more a kind of micromanagement where the project manager tells to the team what to do, and in which time frame.

Our team was not sufficiently involved in the project, and I strongly thought that it was a team's fault. The low team involvement was a problem to the management, and we did our best to solve it.

Once I read "The one minute manager" by Kenneth Blanchard, I found a story that was enlightening. He wrote about a colleague of him, named Mark, that he was not at all involved in his job and one night he went with Mark to a bowling house.

He saw that this colleague was very involved in bowling, and when Mark did a strike, he was very happy, excited, and involved! So he thought "What would happen if in front of the bowling alley there is a curtain and Mark can't see if he strikes or not?". I bet he would not be so involved.

This specific story let me thought about our team involvement. If our team strike can he see it or not? How can I evolve my development process to better involve my team and let them know if they strike? In agile project management, there is a ceremony that meets this goal. It is the iteration review.

This specific ceremony let our team evolve and be the agile team now I'm proud of. During the iteration, the team has the perception of the work done and the approvals form the management.

Furthermore, there is a prize we give to the developer that reaches the higher value point delivery during the year.

Our development team is composed of developer and tester.

Developer develops the stories and set a story ready to be tested. A tester tests the stories.

The developer is in charge of the story produced, so he is the first tester of the story. When he says that a story is a "well-developed" story, he sets a story ready to be tested.

When a story is ready to be tested, the story will be in charge of the tester.

He does first a unit test and then an integration test.

What happen to a story that does not pass the tests? It goes back to the developers!

And what happens to the developers?

We want that the developer has the measure of the works he does, so we set a kind of game between developers.

We give a prize "Developer of the year". Every developer has a score.

The score is the sum of the risk-adjusted costs of the story developed during the year. When a story is developed, the risk-adjusted value of the story is added to the previous score.

But what happens if a ready to be tested story goes back from testing stage to developing stage?

That story will not be added to the developer's score. So the prize is won by the programmer that develops stories that does not come back from testing stage!

Our team today is composed by the **pre-sales chief** that leads the pre-sales engineer, a **post-sales engineer** that is an expert in networking and monitoring and leads the post-sales engineer team, and a development team. Inside the development team, we have a **Chief developer** that is a developer, he writes code, but he acts as a project leader and is in charge of technical decisions.

For each development problem, he talks with the team, but he has the last word. Inside the team, we have technicians specifically skilled in testing (Fig. 1).

The main purpose of the Chief Developer is to create a link between the developing team and the management and reduce the so-called gulf of evaluation.

The "gulf of evaluation" is the difference between what the pre- and post-sales chief (PPSC) are expecting from a story and what the developing team really do!

Our experience says that this gulf could be very large so our Chief Developer has the purpose of understanding their needs and reports it to the developer.

Fig. 1 The team

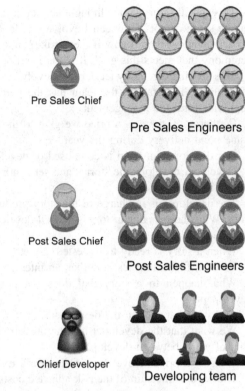

Usually, our Chief Developer, in order to understand whether its perception of the PPSC needs is correct, creates mock-ups. Mock-ups are usually created with power point or HTML and, in our experience, help to drastically reduce the gulf of evaluation.

Our team is collocated. They work on a space that Kent Beck, XP father, calls "caves and commons," that is to say an open space where they works empowering the osmotic communication, but if they need they can go in a "cave" where they can be more concentrate and for a specific job or for private phone calls.

4 The Roadmap

The developing process starts from our roadmap.

Our roadmap is stored in a roadmap board (RMB) that is a classic high-touch low-tech board where we stick the post with epics. A classic backlog!

We tried software solutions, but I felt that they do not have the information radiator performance of a classic white board. Maybe the RMB could be done by a software board because the RMB information radiator attributes are not so important, but we love to work with white board, so we stick with them; furthermore, we strongly have the need to have a very dynamic RMB because the IT market is very dynamic and so are the customer needs and expectations, and nothing is more dynamic than post it on a white board!

How do we fill the RMB and define the roadmap?

There is a committee formed by the chiefs of post-sales and pre-sales (PPSC). The first one has a strict bonding with customers.

He always coordinates a pre-sales activity and has a deep knowledge of the competitor products. He has the vision.

The post-sales engineer is a very skilled technician that is in charge of all the technical activities hold on our product by our customer.

He has a more specific view on daily activities, and he talks with the customer too but at the technician level. They both have the right to create epics and add them to the RMB, but they have to give them a common agreed value point estimation.

So the first ceremony in our iteration project development flow is the RMB refinement. During this meeting, the PPSC and the Chief Developer meet, discuss about new epics or story to add to the RMB, and give them an estimation.

The PPSC are in charge of the RMB, and this is the first rule break from a standard Scrum or agile. In agile, the owner of the backlog is one person.

We have 2, we need this because this 2 persons knows the problems, needs, and wishes of our customers at 2 different level, the pre-sales has a more commercial and strategic view, and the post-sales has a more tactical view.

5 Evaluation

If an epic deserves to be entered in our PB, 3 different evaluations will be given: 1 for the risks, 1 for the costs, and 1 for the added value.

They talk a lot about the evaluations of this 3 values.

The risks evaluated are always technical risks, and we do not talk about social or market risks.

The technical risks can be valued:

- If there are no technical risks: 1;
- If there are a moderate technical risk: 1, 3;
- If the technical risk is high: 1, 8; and
- If the technical risk is very high: No evaluation will be given and a spike must be done in order to solve this risk.

The risk evaluation is always in charge of the Chief Developer.

We use Table 1 to evaluate a technical risks:

Table 1 Evaluate a Technical Risks

Risk evaluation	Situation	Value
No technical risks	We know what to do, how to do and the code produced will not impact previously done features	1
Moderate technical risk	We know what to do, how to do (maybe we have to do some small experiment) and the code produced will have a low impact previously done features	1, 3
High technical risk	We know what to do, how to do (maybe we have to do some small experiment) but the code will have a heavy impact on previously done feature	1, 8
Very high technical risk	We know what to do, but we do not know how	No evaluation

Once evaluated a risk, we evaluate costs. Our measure is in working days, the days that 1 developer needs to do a feature. This is the classical ideal day. The costs evaluation is in charge of the Chief Developer. He really knows his team and how it works.

Once done with the costs evaluation, we multiply the costs value with risks value and we obtain a risk-adjusted cost evaluation.

The value the added value of the product is done using our specific value point definition.

We set to 10 value point the value added to the customer by the multilevel maps (a specific already-developed product feature), and we give a value to the stories comparing them with the value of the multilevel maps.

Now, we can insert the story on the product backlog.

The story with the higher values has the higher priority. Then inside stories with the same value, we give higher priority to the ones that have less risk-adjusted costs.

If a story has very high technical risk, then it cannot have a risk-adjusted cost (as you can see in the previous risk table).

In this specific situation, we value the value point of the story, if the story has less than 3 value point than we put the story at the end of the RMB.

If it has more than 3 value point, we enter the story in the backlog.

When this story will be the highest priority story in the backlog, then we will start a spike.

6 Spike

What is a spike? A spike is a special type of iteration used to mitigate risk and uncertainty in a particular story. Our spike are used to search various technical approaches in the solution domain.

We may use it for the evaluation of potential performance or load impact of a new story, evaluation of specific implementation technology that can be applied to a solution, or for any reason when the team needs to develop a more confident understanding of a desired approach before committing new functionalities.

Spike results are different from iteration results. In iteration results, we produce working code, increments, in a spike the results are information that will be used to resolve the uncertainty that obstacles the identification of a story size.

A spike is strictly timeboxed in one iteration.

At the end of the iteration, we have to stop the spike and give a solution to the problem addressed. If we can do it, we can give a value to the risk and we can consequently give a risk-adjusted cost to the story and add it to the RMB.

We use spike to address the problems or domain where our team have knowledge gaps too. In this specific situation, the team uses an iteration to fill the gaps in order to guarantee the usual delivery speed.

7 Iteration

We uses iterations of 10 days (in Scrum they are called sprints).

The iteration is always started by an iteration planning.

During the planning, the project team chooses the stories to add to the iteration backlog.

The stories are chosen by priority.

During the planning, the Chief Developer explains in detail every story and uses the mock-ups created for the PPSC.

During the meeting, discussions on the stories are encouraged in order to let the team better understand it.

This meeting is timeboxed to 4 h.

Each story is chosen by developers, and they are then in charge of the story. If a specific story needs deeper discussions on behalf of the developers, then a new meeting is set usually at the end of the iteration planning meeting.

Why that?

Because not all the team members need deeper information on that story, the team members that are going to work on other stories can start immediately their own work.

During the iteration, the team member works on their stories and usually they catch up at the end of the week.

So we do not use a ceremony like the daily Scrum.

The reason is that our programmers work on an open space and they have their backlog where stories are split in tasks. When a task is done, the programmer moves it to the DONE zone, so everybody knows what work is done what should be done.

If a programmer needs help because there is some impediment to his job, then he can talk to the Chief Developer.

In this way, our team do not use daily meeting because they catch up constantly during the iteration.

If a programmer ends all his tasks and there is time left during the iteration, then he could help colleagues to end their job or ask to the chief developer if there are other tasks he can do during that iteration.

Usually, we use these time lapse to reduce the technical debt or well-known bugs.

At the end of the iteration, a demo will be done. Our demo involves the development team showing what they have been produced during the ended iteration.

8 Iteration Review

At the end of the iteration, we have the iteration review. The team shows to the PPSC, and the system engineer the increments produced during the iteration.

We do not allow power point slides or other demonstration tool during this event because we want to have a look at the software. The developer team shows the software what it does and how it meets the goal setup for that increment.

This ceremony is used to keep up to date the PPSC that usually, after the planning, is not involved in the developing stage and delegates all the activities to the Chief Developer.

This ceremony is also important for giving the developer team a feedback on the work done. This is very important to us because we want to let them know if they strike or not!

9 Conclusions

Our agile approach to the development process follows the steps previously described.

We know that we are not following by the book any of the agile known framework, XP, Scrum, Crystal, DSDM, or other, but we feel that our development has increased the efficiency and the velocity of the software delivery and creation.

There are ceremonies that we do not have at all like the daily iteration meeting, and there are other that have not a specific space but are carried on during all the developing process like the iteration retrospective.

Our main concern is to try to measure everything in order to try to improve everything. Our measure of the developer efficiency that end with the "Developer of the year" prize is aimed to give to developer a number of their efficiency that helps them to improve the way they work.

The measure of risks or value added by an increment has the same goal.

We know that if you want to improve a process you, have to measure it, and this is what we do!

A New Device for High-Accuracy Measurements of the Hardness Depth Profile in Steels

Roberto Li Voti, Grigore Leahu and Concita Sibilia

Abstract A novel device based on photothermal radiometry is here developed and validated for the NDE and Testing of steels. The use of such an instrument from one side strongly reduces the measurement time of the standard hardness and other mechanical tests, and from the other side allows to performing cheap nondestructive and noninvasive measurements easy to be automatized. A detailed description of both the hardware and the software implemented for the reconstruction of the hardness depth profile will be given in the presentation. The results on gears show the relevant impact in the field of quality control of mechanical components used in many military sectors.

1 Introduction

Quality control of the performance of mechanical components subjected to hardness processing is a topic of fundamental importance, both in the field of automotive and in the field of aerospace systems for both civil and military applications. The lack of cementation, the burns in the steels, the decarburisation of the power gears, and the statoric and rotoric equipment may cause catastrophic failures with serious repercussions. The industry and the companies responsible for the hardening processes as well as for the quality control of the mechanical components are continuously seeking for the improvements in the standard destructive tests performed by Vicker or Brinell durometer where one mechanical component is chosen for random testing.

Since 1996, the use of IR systems based on photothermal radiometry for the nondestructive determination of the hardness profiles in steels has been deeply studied and discussed by several groups between Europe in the framework of European Thematic Networks (BRRT-CT97-5032) [1], North America [2], and more recently Asia [3] as shown by the huge numbers of papers in the field.

R.L. Voti (✉) · G. Leahu · C. Sibilia
Dipartimento S.B.A.I. Sez. Fisica, Sapienza Università di Roma, Via A. Scarpa 16, 00161 Rome, Italy
e-mail: roberto.livoti@uniroma1.it

© Springer International Publishing Switzerland 2016
P. Ciancarini et al. (eds.), *Proceedings of 4th International Conference in Software Engineering for Defence Applications*, Advances in Intelligent Systems and Computing 422, DOI 10.1007/978-3-319-27896-4_20

In recent years, several technical problems have been solved to improve the accuracy in the reconstruction: Some of the crucial issues were related to the following specific parts of the PTR system:

(a) Pumping system: absorbed power, pump wavelength, optimum pump spot-size, and related modeling of the 3D heat diffusion effects. Time, frequency, or chirped regime.
(b) IR detection system: detection on/off axis with respect to the pumping axis, spectral response.
(c) Accuracy in the calibration with reference hardened sample in order to establish an accurate anticorrelation curve between thermal diffusivity and hardness for the specific hardened steel.
(d) Samples: roughness, size, and geometry (flat or curve) of the samples for industrial applications.
(e) Inverse methods for the retrieval of the depth profile: Least square fitting, Polynomial approximation, Neural Networks, Singular Value Decomposition, Genetic Algorithms, etc.

Obviously the global accuracy is strictly dependent on the accuracy of each specific part.

2   New Approach for NDE and Testing

In this paper, we introduce a new light and compact PTR device for a high-accuracy measurement of the cementation in gears based on photothermal depth profiling [4, 5].

The hardware of this system has been improved, made more compact, and integrated with mechanized and robotic arms for industrial needs (see Fig. 1).

Fig. 1 PTR compact device for 3D measurements of the cementation in gears

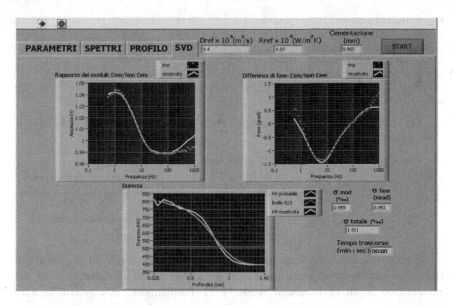

Fig. 2 New software for the reconstruction of the hardness depth profiles in AISI9310 gears

In particular, the system uses a powerful IR laser working at 9 W, modulated at a frequency ranging from 1 Hz to 4 kHz so to generate the thermal waves with a penetration depth from microns to millimeters. The waves penetrate the surface layer and are reflected from the inner martensite/austenite interface. The effects of thermal wave reflection are revealed by a Peltier cooled infrared sensor working in the range MIR 3–5 μm [6].

The software of the system has been developed so to retrieve the diffusivity/hardness depth profile by applying the Singular Value Decomposition tool to solve the severely ill-posed inverse problem of the photothermal depth profiling. The SVD is usually truncated ad hoc so to limit the effect of the noise with some minimal loss of spatial resolution. This innovative software has been implemented in Labview, so to be run in real time with the data acquisition (see Fig. 2).

3 Experimental Results

Preliminary results on both AISI 9310 and Pyrowear 53 hardened steel gears show accurate hardness profile reconstructions in comparison with the hardness measurements by standard Vicker test. In Fig. 3, for example, the hardness depth profile determined nondestructively by using the PTR (blue line) is compared with the one measured destructively, a posteriori, by Vicker test (red line). The excellent agreement between PTR and Vicker measurements has been confirmed on the whole set of samples, where the error in the estimate of the effective cementation depth is usually below 0.1 mm.

Fig. 3 Hardened depth profiles in AISI9310 steels. Comparison between standard Vicker test (*red line*) and the new PTR nondestructive test (*blue line*) (Color figure online)

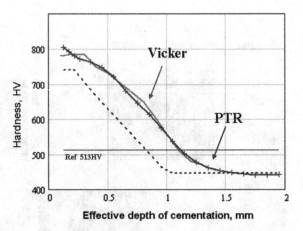

In synthesis, we have here established the fundamental role of photothermal radiometry as nondestructive technique applicable to the determination of the cementation depth of hardened steels [4–8]. Furthermore, this system is very flexible and can be applied also for the determination of the thickness of coatings and/or paints, and for the detection of cracks and defects, when thermal waves with a different range of frequency are used. The future idea is then to extend the applications of photothermal radiometry to other mechanical components such as those covered with protective coatings such as paints and/or ceramic tiles.

References

1. Walther HG, Fournier D, Krapez JC, Luukkala M, Schmitz B, Sibilia C, Stamm H, Thoen J (2001) Photothermal steel hardness measurements-results and perspectives. Anal Sci 17:s165–s168
2. Munidasa M, Funak F, Mandelis A (1998) Application of a generalized methodology for quantitative thermal diffusivity depth profile reconstruction in manufactured inhomogeneous steel-based materials. J Appl Phys 83:3495–3498
3. Wang C, Mandelis A, Qu H, Chen Z (2008) Influence of laser beam size on measurement sensitivity of thermophysical property gradients in layered structures using thermal-wave techniques. J Appl Phys 103:043510
4. Glorieux C, Voti RL, Thoen J, Bertolotti M, Sibilia C (1999) Depth profiling of thermally inhomogeneous materials by neural network recognition of photothermal time domain data. J Appl Phys 85(10):7059–7063
5. Glorieux C, Voti RL, Thoen J, Bertolotti M, Sibilia C (1999) Photothermal depth profiling: analysis of reconstruction errors. Inverse Problems 15(5):1149–1163
6. Li Voti R, Liakhou GL, Paoloni S, Sibilia C, Bertolotti M (2001) Thermal wave physics. J Optoelectron Adv Mater 3:779–816
7. Li Voti R, Leahu G, Gaetani S, Sibilia C, Violante V, Castagna E, Bertolotti M (2009) J Opt Soc Am B 26, 1585–1593
8. Dehoux T, Wright OB, Li Voti R, Gusev VE (2009) Phys. Rev. B 80, 235409

AGILE Methodology in Progesi MDA Model (Meta–Dynamic–Agile)

Marco De Angelis and Roberto Bizzoni

Abstract Contrary to popular belief, CMMI and Agile cannot only coexist forcibly in a company, but also live peacefully together which benefits from their integration. This paper reports the experience of Progesi, a company in the market of defense and public administration that has adopted the method dev CMMI level 3 and uses daily methodology in Agile projects. Examples of integration are available in the literature, and our analysis starts from some well-known case study: In fact, many have had the need to find a meeting point between the two philosophies. The key to integration is an approach that we called "Meta–Dynamic–Agile" model, which is based on "competence centers."

1 Introduction

This article discusses, briefly, as you can manage projects with the Agile methodology in a company which adopts CMMI processes.

In general, we found that the concepts of Agile can provide solutions to the demands of the CMMI, simplifying, Agile can provide the "how to" to the CMMI demands of "what" and in this connotation the two methods are not contradictory.

In this article, we want to focus on the fact that CMMI and Agile are "methodologies," and therefore, they do not provide solutions for use by any organization in any context, but a set of guides and tools that the organization must necessarily contextualize into their own evolving environmental situation.

M. De Angelis (✉) · R. Bizzoni
Progesi, Roma, Italy
e-mail: mdeangelis@progesi.it

R. Bizzoni
e-mail: rbizzoni@progesi.it

© Springer International Publishing Switzerland 2016
P. Ciancarini et al. (eds.), *Proceedings of 4th International Conference in Software Engineering for Defence Applications*, Advances in Intelligent Systems and Computing 422, DOI 10.1007/978-3-319-27896-4_21

Progesi has developed the concept of "centers of competence" (*centri di competenza*), which are characterized by:

- Projects are grouped into clusters according to the kind of expertise necessary to achieve their objectives (domain-oriented and technological-oriented).

  The managers (leaders) of the competence centers ensure a strong link between the projects (production) and the organization (management control, organizational services, business units …).

  The leaders and representatives of the organization meet periodically in a board.

  We called "meta" this characteristic of being a group of representatives of groups of projects.
- The competence centers are not static in time, but represent clusters of projects and therefore vary depending on the business situation, for instance at this moment at Progesi are active five competence centers.

  This is why our scheme is called "dynamic."
- The competence centers operates internally according to the Agile logic to help continuous improvement, both for the organization and for the customer.

This approach to the work has been developed over time, and it is evolving. By virtue of the three properties listed, it creates a methodology that we called MDA (meta–dynamic–agile).

2 Background: Progesi

Progesi is a company that works in the "Defense and Space" and "Civil" (government, telecommunications) markets that has been certified CMMI for Development level 3 [1] in 2013.

In some projects, the customer or our partner has chosen the use of Agile practice, in particular the Scrum method.

CMMI processes are agnostic about the methodology; they have focused on the "what" and not on the "how to," so we accepted the demands and we sought to use the best characteristics of the two methodologies.

Obviously, in the managing of mixed teams (i.e., Progesi and/or customer and/or other companies) are concentrated the greatest difficulties, and soon enough, we met some resistance among the CMMI-oriented groups and the Agile-oriented groups.

We carried out an analysis to understand the roots of this communication troubles; the results are as follows [2]:

- The main reason seems to be the lack of knowledge of the respective methods; the Agile groups perceived CMMI processes as heavy, long, tedious, and often unnecessary; for the CMMI group, the Agile methodology is often inaccurate, uncontrolled, and unpredictable.

– The approach to projects CMMI is top-down and is driven by the organization, while the agile approach is bottom-up and is driven by subject matter experts; this approach is also true as regards the improvement phase.
– CMMI groups tend to deal with mission-critical projects with high risks and with an OBS (organization breakdown structure) very structured, while Agile groups tend to deal with smaller projects, with tight deadlines and instable requirements.

The fundamental error is to perceive the methodologies as standards to guide the project team in terms of "what to do" and "how to do," while the truth is that they drive the organization toward improvement of its business without rigid constraints [3].

The goal was to allow—in the most natural way possible—the managing of projects with Agile methodology together with our partner within the organization.

3 What Is Agile

The agile concept is older than often believed; it has taken the first steps since the 1950s.

In 1976, Tom Gilb wrote the book "Software Metrics" [4] in favor of agile development based on adaptive iterations that provided rapid results and more business benefits.

This approach becomes more mature in the book wrote by Barry Boehm "The Spiral Model of Software Development and Enhancement" [5].

During the 1990s, the agile concepts have been extended and various methods have been theorized and applied in many companies; some examples are as follows [6]:

– rapid prototyping,
– Rapid application development (RAD),
– Rational Unified Process (RUP),
– Extreme programming (XP),
– Feature-driven development (FDD).

The need to coordinate and to develop the Agile methods has led a group of experts in the creation of the Manifesto for Agile Software Development [7]. Some of authors of the manifest formed the "Agile Alliance" a nonprofit organization dedicated to promoting the adoption of agile methods.

Some of the original authors of the Manifest (and others) have established a set of six management principles known as the "Project Management Declaration of Interdependence" (DoI); the Manifest, the Interdependence (DoI) publications and the not-profit organizations created to promote the Agile approach have resulted in the rapid growth and adoption of Agile methods [8] .

Some methods, most notably Scrum, continue their grow in the software industry.

3.1 The Scrum Method

The method Scrum is an agile method in that it is born to meet the challenges of a rapidly developing and changing requirements; the ability to respond to change is one of its strengths [9].

Scrum has been described by Ken Schwaber [10, 11] in 1996 and is based on the fact that the development process is not predictable, the requirements are not known in detail at the beginning of development, and customer needs can change over time, introducing the "do what it takes."

The iterative process and incremental process allow controlling the variables of the project (time, requirements, resources, tools, etc.) that cannot be planned with precision at the beginning.

Scrum is based on the following idea: implementing an iterative and incremental skeleton through three roles: the product owner, the team, and the scrum master.

Role	Responsibilities
Product owner	Represents the interests of the stakeholders and maintains the prioritized list of requirements (product backlog)
Team	Teams are self-managing, self-organizing, and cross-functional, and they are responsible for figuring out how to turn product backlog into an increment of functionality within an iteration and managing their own work to do so
	Team members are collectively responsible for the success of each iteration and of the project as a whole
Scrum masterr	Responsible for managing the Scrum process, i.e., for teaching Scrum to everyone involved in the project, for implementing Scrum so that it fits within an organization's culture and still delivers the expected benefits, and for ensuring that everyone follows Scrum rules and practices

The project carried out according to the Scrum methodology starts from a high-level view of the system; subsequently, the participants created a list of requirements (product backlog) divided by priority in iterations called sprints.

A Sprint is a period of 30 days (generally) of development, it begins with a planning meeting with the product owner and the Team, they select from product backlog the high-priority requirements, and then, the product owner explains what he needs and the Team what it is able to develop in the next Sprint [12].

At the end of the meeting, they fill a list of tasks to complete the sprint (Sprint Backlog).

During the sprint, the team meets daily in meetings of 15 min to monitor the progress of the work; the team answers three questions as follows:

- What activities have you done since the last meeting?
- What activities will you do before the next meeting?
- Do you have any obstacles?

At the end of the Sprint, in the Sprint Review meeting the Team presents what has been developed during the Sprint to the product owner and all other stakeholders present, after the scrum master holds a meeting to enhance the next Sprint.

4 What Is CMMI

According to the Software Engineering Institute (SEI), the CMMI Capability Maturity Model Integration (CMMI) is a model of maturity of process improvement for the development of products and services. It consists of best practices that relate to the development and maintenance activities that cover the life cycle of the product, from conception through delivery and maintenance.

In CMMI, the central theme is the process management. In this context, process refers to "what to do" rather than to "how to do it." The process communicates how the organization expects the work is done, what are the objectives, what are the roles, and which tools are available.

The process is a tool to increase business, through transparency and measurement of work done that provides the necessary information to make decisions that guide the development of the product.

CMMI describes 22 process areas. A process area is a group of related activities that are performed collectively to achieve a set of goals.

A process area goal describes the result expected and the activities to achieve those goals.

Some goals and practices are specific to the process area; others are generic and apply across all process areas: They describe the essential ways in order to institutionalize a process. Institutionalization refers to a process's degree of repeatability, standardization, and sophistication of control.

CMMI does not describe a single process nor a standard, and then, each organization could achieve the goals in a different way; the processes are effective if:

1. They are distributed in all projects;
2. They are reviewed and improved periodically;
3. They are compatible with practices in use and flexible; and
4. They are understood throughout the organization by appropriate training.

The processes are aimed to create trust and therefore to achieve a balance between the needs of the organization and the needs of projects and professionals.

4.1 A Model, Not a Process Standard

CMMI does not provide a single process, rather the reference models for how to proceed to improve processes, because it was designed to compare organization's existing processes with best practices developed by members of industry, government, and academia; reveal areas for improvement; and provide ways to measure progress.

CMMI is a model, not a standard that requires results with little variation from one implementation to another; it contains neither processes nor procedures.

The practices are intended to encourage the organization to experiment and use other approaches, models, frameworks, practices, and process improvement, as appropriate, based on the needs and priorities of the organization.

CMMI is a model, that is, an ideal to which the organization can learn and refers to real situations, so that has to be implemented and not applied as a further practice on existing business practices.

When using CMMI as a model for process improvement, rather than a standard for this, organizations are free to implement any application of process improvement and meet their needs, it could be possible to see the progress in the organization beyond the project level and not only the immediate projects and developers.

5   Comparing CMMI and Agile

In order to compare the two models for proceeding with a merger, or at least to ensure their "peaceful coexistence," we compared some key aspects, making help from CMMI-oriented and Agile-oriented groups to test the strengths of both [13].

From a methodological point of view, we made a gap analysis in the project management areas that highlights the strengths, weaknesses, and recommendations that can be used to improve the implementation of the model [14].

CMMI is structured in 22 areas ordered by category of area and level of maturity; in this comparison, the project management areas until level 3 are been considered (except for SAM).

Process area	Category	Maturity level
Causal analysis and resolution (CAR)	Process area support	5
Configuration management (CM)	Support	2
Decision analysis and resolution (DAR)	Support	3
Integrated project management (IPM)	*Project management*	*3*
Measurement and analysis (MA)	Support	2
Organizational process definition (OPD)	Process management	3
Organizational process focus (OPF)	Process management	3
Organizational performance management (OPM)	Process management	5
Organizational process performance (OPP)	Process management	4
Organizational training (OT)	Process management	3
Product integration (PI)	Engineering	3
Project monitoring and control (PMC)	*Project management*	*2*
Project planning (PP)	*Project management*	*2*

(continued)

(continued)

Process area	Category	Maturity level
Process and product quality assurance (PPQA)	Support	2
Quantitative project management (QPM)	Project management	4
Requirements development (RD)	Engineering	3
Requirements management (REQM)	*Project management*	*2*
Risk management (RSKM)	*Process management*	*3*
Supplier agreement management (SAM)	Process management	2
Technical solution (TS)	Engineering	3
Validation (VAL)	Engineering	3
Verification (VER)	Engineering	3

Process Area: Project Planning

The purpose of Project Planning (PP) is to establish and maintain plans that define project activities; it has 3 specific goals (SG 1 establish estimates, SG 2 develop a project plan, and SG 3 obtain commitment to the plan); the mapping of the found gap—with reference to specific practice—is presented below [15] .

Specific goals	Gap (specific practices)
Establish estimates	No method is explicitly mentioned to guide the estimates (*establish estimates of work product and task attributes*)
	Cost is not explicitly mentioned (*estimates effort and cost*)
Develop a project plan	Scrum does not provide orientations about establishing budget (*establish the budget and schedule*)
	The risk identification occurs during daily meetings, without using organization categories and source (*identify project risks*)
	There is not a formal procedure to collect, consolidate, and publish project information to manage data privacy (*plan data management*)
	Scrum team is composed of skilled people; it does not mention the necessity to planning the knowledge and skills needed to perform the project activities (*plan needed knowledge and skills*)
Obtain commitment to the plan	–

Process Area: Project Monitoring and Control

The purpose of project monitoring and control (PMC) is to provide an understanding of the project's progress so that appropriate corrective actions can be taken when the project's performance deviates significantly from the plan.

PMC encloses 2 specific goals (SG 1 monitor project against plans and SG 2 manage corrective action to closure).

Specific goals	Gap (specific practices)
Monitor project against plans	Team monitors the project in the daily meeting, but the cost, size, and effort estimates are not carried out in a systematic manner; there is not a formal tracking of them as required by CMMI model (*monitor project planning parameters*)
	Risks are tracked in an informal way (*monitor project risks*)
	Scrum does not adopt any procedure for planning and tracking data management (*monitor data management*)
Manage corrective action to closure	Corrective actions are taken for the impediments found. However, there is not any register of how these actions are planned and monitored (*take corrective action*); the results of corrective actions are not analyzed to determine its efficacy (*manage corrective action*)

Process Area: Integrated Project Management

The purpose of integrated project management (IPM) is to establish and manage the project and the involvement of relevant stakeholders according to an integrated and defined process that is tailored from the organization's set of standard processes.

IPM is composed of 2 specific goals (SG 1 use the project's defined process and SG 2 coordinate and collaborate with relevant stakeholders).

Specific goals	Gap (Specific practices)
Use the project's defined process	Scrum does not define a set of organizational standard processes, but just for the project (*establish the project's defined process*)
	Scrum does not contribute to the asset of the organization (*contribute to organizational process assets*)
Coordinate and collaborate with relevant stakeholders	Scrum, dependencies, and risks can be managed as impediments, being identified through Scrum Daily Meetings. Scrum master is in charge to resolve any identified problem as soon as possible, so dependencies as well. However, there are not registers of negotiation, meeting minutes, or agreed dates to remove such dependencies (*manage dependencies and resolve coordination issues*)

Process Area: Risk Management

The purpose of risk management (RSKM) is to identify potential problems before they occur so that risk handling activities can be planned and invoked as needed across the life of the product or project to mitigate adverse impacts on achieving objectives.

RSKM is composed of 3 specific goals (SG 1 prepare for risk management, SG 2 identify and analyze risks, and SG 3 mitigate risks).

Specific goals	Gap (specific practices)
Prepare for risk management	Scrum does not mention practices to define sources, parameters, or categories (*determine risk sources and categories and define risk parameters*)
Identify and analyze risks	In Scrum, risks are identified, but the assessment, categorization, and priorization of these risks occur in an informal manner (*evaluate, categorize, and prioritize risks*)
Mitigate risks	In Scrum, there are not strategies to risk response or a mitigation plan for the critical risks based on historical bases or something like that (*develop risk mitigation plans*)

Process Area: Requirements management

The purpose of requirements management (REQM) is to manage requirements of the project's products and product components and to ensure alignment between those requirements and the project's plans and work products.

REQM is composed of a single specific goal (SG1 manage requirements).

Specific goals	Gap (specific practices)
Manage requirements	*Scrum handles requirement, but* there is not a formal tracking of them as required by CMMI model (*maintain bidirectional traceability of requirements*)

5.1 Gap Analysis from CMMI Perspective

The goal of this gap analysis was to compare the agile method Scrum in relation to the project management process areas of the CMMI model, showing the gap and the strength existents between them. From this mapping, we conclude that Scrum does not cover all the specific practices of the project management process area, but it could be tailored to be more compliant with CMMI, and on the other hand, a CMMI organization has already implemented the practices not covered.

It is essential to find an agile mechanism that allows the organization to transmit the practices to projects managed by Scrum methodology. Obviously, this mechanism, which in this case consists of the "centers of competence," cannot change the methodology Scrum.

5.2 Criticism from Agile Perspective

Here, it is interesting to quote some of the criticisms that the groups have advanced toward the agile CMMI.

Some of these criticisms relate to the implementation of the CMMI and not to the model itself; however, we feel very useful because these criticisms have improved some aspects of the process, especially with regard to the formation and institutionalization of the process [16].

CMMI criticism	Scrum support
CMMI is focused on process rather than people	Scrum treats people issues, the same as other project management issues
CMMI does not focus on underlying organizational problems	Scrum Master maintains an impediment list and resolves organizational, personal, and technical issues
The quality of the process does not guarantee the quality of the deliverables	Product owner is responsible for the quality of the deliverables, while Scrum Master is responsible for the quality of the process
CMMI ignores technical and organizational infrastructures	In the Scrum meetings, the team focuses on technical and organizational issues

5.3 Analysis Result

The result of the analysis is that CMMI and Agile are compatible.

At the project level, CMMI works on a high level of abstraction and focuses on what practices implement and not on the development methodology adopted, while agile methods focus on how projects develop products [17].

Therefore, CMMI and agile methods can coexist, although it seems more natural to introduce methodologies agile in CMMI contexts rather than vice versa; CMMI and Agile can complement each other, creating synergies for the benefit of the organization using them.

Agile methods generally do not provide practices and guidelines for the implementation of an Agile approach throughout the organization, while CMMI supports a long-term process of adaptation and improvement across the company as a whole.

6 Our Solution: Competence Centers Board

The idea is to classify the projects depending on the competence required into clusters called Competence Centers ("centri di competenza").

Competence is a term concerning: skill, knowledge, qualification, and capacity, so it is very concrete and related to people.

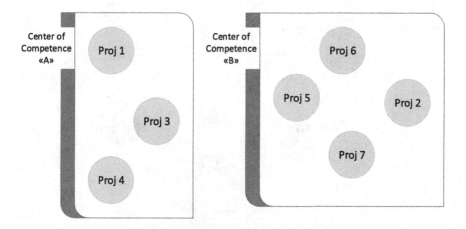

The competence centers are deemed by the organization of the lines along which develops the business; examples could be:

- VoIP Systems;
- Automotive Systems;
- Command and Control Systems; and
- Tracking Systems

The competence centers aim to deepen and widen strategic competence and increase the synergy between projects.

If a new project is not within a competence centers, it means that a new potential market can be born; if the organization believes it is in line with the business strategic view, then a new center of competence is created with a single project.

Each cluster of projects has a leader who is a subject matter expert, and who knows and follows closely the projects and supports the organization (BU managers, management control, quality manager, etc.) in business and in bringing the organization objectives and business processes at the project level.

The leaders of competence centers and the organization constitute the Competence Centers Board (hereafter Board); it is an essential tool for improving the performance of the organization.

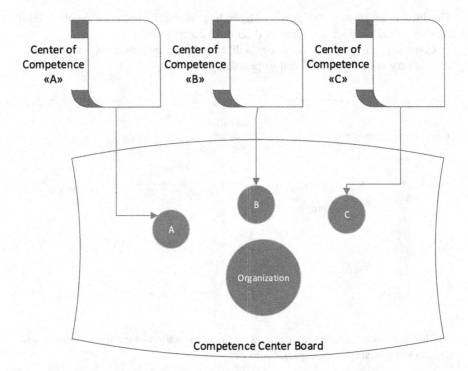

- The board is particularly light and quick, because leaders of the CC actively participate in the projects, have a very deep understanding of the technologies and application domains, and represent a natural link between organization and projects.

The board meetings are periodic, very fast, and concrete; the agenda consists mainly of two questions:

- How can the organizations support projects?
- How can the projects support the organization?

Leaders of the competence centers respond to the first question; they report the activities going on, their needs, and their ideas to improve the quality of deliverables.

The organization (usually Business Unit Managers) respond to the second question, what are the reporting lines of business, the demands of the company, and the new market opportunities.

Requests and proposals are evaluated and discussed in the board; the board adopts an agile approach through an iterative "agile unit" (described below) and implements what is commonly decided.

In this context, the single project within a cluster can be well managed by Scrum methodology, and gap analysis in fact shows that the organization can provide a set of tools and services to better support the projects.

For example, regarding the risks, the competence centers use the RBS corporate and tables of corporate priority.

During the daily meeting, the team follows the methodology Scrum using tools provided by the organization and the lessons learned are transmitted to the organization by the leader of the competence centers.

From this example, it is evident how the team should not change habits or burden the function and indeed has available tool that helps people and support for improving the work.

The role of leader of the competence centers is different from Scrum Master role, they are not overlaid in any case, the two work together because the scrum master focuses on the project while the leader provides link among the organization and the projects.

The following paragraphs will show the three characteristics that give the name to the methodology: agile, dynamic, and meta.

6.1   Agile Approach

The board meets regularly; it is not necessary that the meeting is colocated; often meetings take place via Skype. The meetings are managed with an agile approach (unit agile) which increases the trust relationships between projects and organization.

Unit agile means an iterative process of four steps:

Design

The presentation of a need, an idea, a tool, or a process is the step in which you create a proposal to be discussed.

Comparison with partners

When the proposal is agreed with the other leaders of center of competence and the organization, this step analyzes the criticism, considering the change requests.

Implementation

The proposal is developed; it may be a tool, a process, a software component, a document, presentation, and so on.

Test

It is time for verification and validation of what has been achieved with the partners and will lay the groundwork for the new iteration.

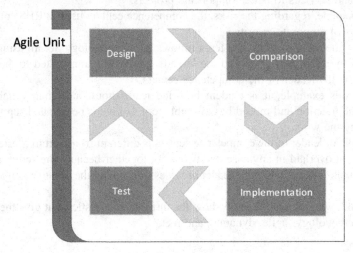

A key part of the AU is the comparison because at that stage is reached the commitment, and all the stakeholders of the company are present or represented at the highest level.

The agile unit is repeated as many as it is necessary to achieve the objective. This is an iterative approach and is in very close contact with customers and organization.

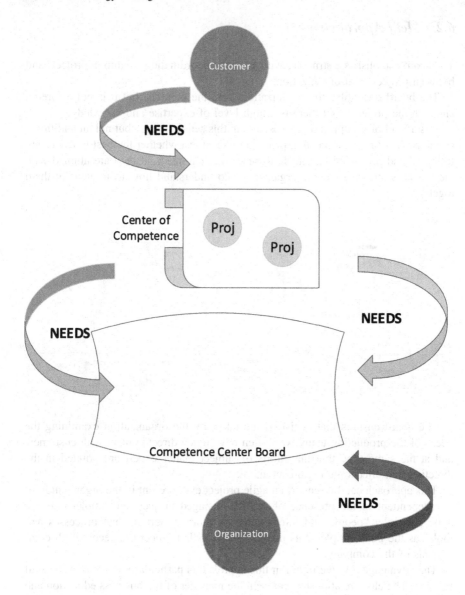

Customers interact with project teams, competence centers leaders collect needs and aggregate them for the board, competence centers have internal needs (processes, needs, tools, etc.), and organization has business strategy to be achieved, in a continuously and iterative methods.

6.2 Meta Approach

The board establishes a structure with a regular coordination rhythm in projects and between projects and organization.

The board decouples the organization from the projects, but it is not a foreign entity in the projects, but there is a high level of expertise and knowledge.

It is a kind of grouping of groups, and in this sense "meta," but not an additional structure; it is instead a meeting place in order to test whether the best practices are followed and improved and made to serve the real needs, if they are aligned with the business strategies of the organization to understand how to implement them together.

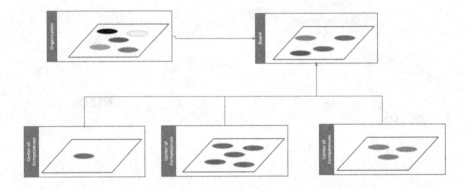

The board ensures that decisions are taken by the organization examining the needs of the production team, who often have much direct contact with customers and at the same time that the efforts of subject matter expert are pointed in the direction that affects the organization.

This approach can be seen as an agile project ever present in the organization; it is independent of the presence of AGILE-managed projects and implements the agile principle: "people and interactions are more important than processes and tools" as the board implements processes and tools to meet the needs of all components of the company.

The advantage that we found in this approach is particularly evident in the staff training. The close relationship between the manager of the business education and the project managers leads to a continuous production of courses that are designed, annotated, and implemented in different levels.

6.3 Dynamic Approach

The board is the company: both for organization and for projects sorted by skills: A key feature of the board is that it changes over time for two main reasons:

- Change the cluster of projects: As the market changes, change the growth opportunities and skills mature.
- Change the organization, because the business strategies change the composition of the organization involving new actors and new roles.

The board reflects the composition of the company and therefore changes over time; it adapts to the needs and the market.

If a cluster of projects becomes empty, there may be no need of the competence centers and the leader does not participate in the board.

Further, the CMMI promotes the continuous improvement of the company—the IDEAL cycle—that cannot be achieved by a static and formal configuration.

In this dynamic approach, the defined processes of the organization guarantee the interactions between the projects through leaders of centers of competence and increase mutual trust.

The processes are improving organization and to be adapted to market (customers) needs; with this approach, it is possible to meet the different needs of the stakeholders and can combine quality and speed of intervention that are fundamental peculiarities in the market.

7 Conclusion

In general, the agile methodology can be used in organizations CMMI-compliant, and there are many examples; in this article, we presented the solution developed (or rather under continuous development and improvement) in Progesi.

We presented the MDA methodology, which actually is not a real methodology rather is an implementation of the CMMI method in the presence of a board operated with the agile.

For simplicity, we called this particular implementation methodology MDA, and the acronym presents 3 fundamental characteristics of the board: meta, dynamic, and agile.

The MDA focuses on a division of the projects in clusters sorted by competence and the creation of a board that brings together leaders of these competence centers and the reality of the organization that the company considers essential to pursue the current business strategy.

Within the clusters, the leaders of competence centers support projects, also agile, in the best practices identified by the organization, and collect the needs of the clients and of the teams.

The board is a synthesis between different needs:

- teams in the projects;
- organization;
- customers (reported by teams);
- customers (reported by the organization).

This synthesis occurs through an agile approach in 4 phases:

- The presentation of a need and ideas for resolution
- Commitment from organization and other competence centers
- Implementation of solution
- Verification of the solution

At each periodic meeting (typically monthly), needs are collected, and solution is implemented and monitored, in a very dynamic and agile way.

The methodology is constantly evolving; it will be interesting to analyze the contribution to the company over time and his ability to respond to market challenges.

References

1. CMMI Product Team (2010) CMMI® for development, version 1.3 CMMI-DEV, V1.3", SEI, Nov 2010
2. McMahon, PE (2012) PEM systems. Taking an Agile organization to higher CMMI maturity. CrossTalk—Jan/Feb 2012
3. Glazer H, Dalton J, Anderson D, Konrad M, Shrum S (2008) CMMI® or Agile: why not embrace both! SEI, Nov 2008
4. Gilb T (1977) Software Metrics. Winthrop Publishers
5. Thayer RH, Boehm BW (1988) Software engineering project management. Computer Society Press of the IEEE
6. Highsmith J (2004) Agile project management, creating innovative products. Addison-Wesley, Reading
7. AgileManifesto (2006) Manifesto for Agile software development, http://agilemanifesto.org, Dec 2006
8. Highsmith J (2002) Agile software development ecosystems. Addison-Wesley, Boston
9. Cochango (2006) Scrum for team systems. http://www.scrumforteamsystem.com, Dec 2006
10. Schwaber K (2004) Agile project management with Scrum. Microsoft
11. Schwaber K (2006) Controlled chaos: living on the edge. http://www.controlchaos.com/oldsite/ap.htm, Dec 2006
12. Larsen D (2006) Agile retrospectives. Pragmatic Bookshelf
13. Hansen MT (2009) BestBrains "From CMMI and isolation to Scrum, Agile, lean and collaboration". In: Agile conference
14. Marçal ASC, de Freitas BCC, Furtado Soares FS, Belchior AD (2007) Mapping CMMI project management process areas to SCRUM practices. Software engineering workshop, 2007. SEW 2007. 31st IEEE, Mar 2007
15. Cohn M (2005) Agile estimation and planning. Prentice Hall, Englewood Cliffs
16. Potter N, Salkry M (2011) Implementing Scrum (Agile) and CMMI together. Scrum Alliance (web site), Feb 2011
17. McMahon PE (2011) Integrating CMMI and Agile development: case studies and proven techniques for faster performance improvement. Addison-Wesley, Reading

Data Breaches, Data Leaks, Web Defacements: Why Secure Coding Is Important

Raoul Chiesa and Marco De Luca Saggese

Abstract On last December 2014, Security brokers (SB)—"Targeted Threats Team"—successfully completed a full analysis, which started back on January 2013, operating over 24 months of deep research and data correlation. Analyzing those main and biggest security incidents and data breaches occurred over the last ten years, starting from the faraway year 2004. The lesson learned was quite impressive and may somehow shake the modus operandi and the mental approach we are used to. This paper aims to recap those key points emerged from that research project, and those new logics we should internally apply within our organizations over the next months and upcoming years. This paper aims to provide, on its first section, the so-called big picture toward those main threats linked with information theft and leaks, and Web defacements, along with those consequent impacts on organizations, through keywords such as Cyber Intelligence, both from open and from closed sources. The second section of the paper provides a general overview of the importance of the so-called secure programming and on those typical mistakes that pop-up when running security testing projects, and advanced penetration testing activities, toward Web applications.

1 Introduction

The idea of a joint research study popped up over the last three months of 2014, while we were analyzing those data as reported by the "World's biggest data breach" [1]: ten years of data breaches, data leaks, and data thefts, launched against all of the possible market's sectors (as reported in Table 1).

R. Chiesa
Security Brokers SCpA, Milano, Italy
e-mail: rc@security-brokers.com

M. De Luca Saggese (✉)
Italian Army, Rome, Italy
e-mail: marco.delucasaggese@esercito.difesa.it

© Springer International Publishing Switzerland 2016
P. Ciancarini et al. (eds.), *Proceedings of 4th International Conference in Software Engineering for Defence Applications*, Advances in Intelligent Systems and Computing 422, DOI 10.1007/978-3-319-27896-4_22

Table 1 Data breach possible market's sectors

✓ Academic	✓ Military
✓ Energy	✓ Retail
✓ Financial	✓ Tech
✓ Gaming	✓ Telecoms
✓ Government	✓ Transportation
✓ Health care	✓ Web companies
✓ Media	

Alone, the list itself would be enough in order to realize the dimension and size of the problem. Reported authors organized the "leak methodology" into the following criteria: incident publishing hacked, insider, lost/stolen laptop, lost/stolen digital media, and poor security.

While reading the report, all of us were definitely astonished; our memory ran back to February 2014, when we analyzed those (excellent!) "Strategies to mitigate cyber intrusions" published by Australian Signals Directorate (ASD), from the Department of Defense of Australia: Yes, we are speaking about those very well educated and skilled team, which is used to investigate (and solve) hacking incidents since the 1990s [2].

While browsing through the guidelines published by the ASD, we were surprised while learning that no explicit rules or advices have been given regarding secure coding or secure programming. We have found "User application configuration hardening," in order to address intrusions that exploit the prevalence of Java vulnerabilities or involve malicious macro-code in the Microsoft Office files, and Application whitelisting (#1 in the ASD list), but still not a specific item, related, i.e., to S-SLDC (secure software life development cycle), OWASP Top-Ten, or general warnings and best practices on how to program securely. And, as you will read on your own, no mention of data feeds from the cyber crime intelligence environments.

That is why along with SB team we decided to run a different analysis, highlighting those macro-errors, both procedural and technological, which rise up from this impressive amount of data, helping out the readers and the audience to understand how much two keywords might provide an higher layer of defense, preventing such unpleasant scenarios: Cyber Intelligence and Secure Coding.

The first evidence that pops up is quite scary: No one of the victim's organizations had a single idea of "who or what" hit them, despite the very important budgets they spent for a "better" IT security, products, software, and top consultants. This is quite significant and brings us to a paradigm, which will gain major importance along the upcoming years. As of today, the information about breaches to our companies and organizations, as well as those *signals* and indicators of next, upcoming targeted attacks, must be found *outside* of our environments. We should then talk about Cyber Intelligence, meaning *Intelligence* ("information and data") applied to the cyber context, and to cyberspace, in which we do all operate every day.

2 Cyber Intelligence (from Open Sources)

Cyber Intelligence is a service that can be applied toward all different types of business, ranging from private enterprises (finance, energy, water, fashion, gaming, etc.) up to government and military contexts. Nowadays, it is mandatory being always updated on what is happening, in order to better understand what may happen—and will, certainly, happen.

Such services are based on two different approaches: open-source-based and closed-source-based intelligence. In the first one, specialized companies crawl from the world's Internet space (IPv4, IPv6, Web sites, news, etc.) all of the available information, and organized them by sector, typology, and keyword. An example is provided below (in the following "Fig. 1"), as a screenshot took from the Web portal "BRICA" [3].

As we can see, data and information are matched with the customer's profile, and based specifically on the business sector and IT infrastructure of the customer itself.

In this specific approach, BRICA does not manage and organize only those security alerts and security news which are "technology-related" but, instead, collects, analyzes, and classifies "Early Warning type" contents, which are carefully selected, and related to different types of cyber risks and cyber threats, for different business categories. The right information on new global threats, emerging malwares, and cybercrime campaigns, frauds and hoaxes, environmental threats, and hacker's activities, as well as terrorism (local and global) and human health.

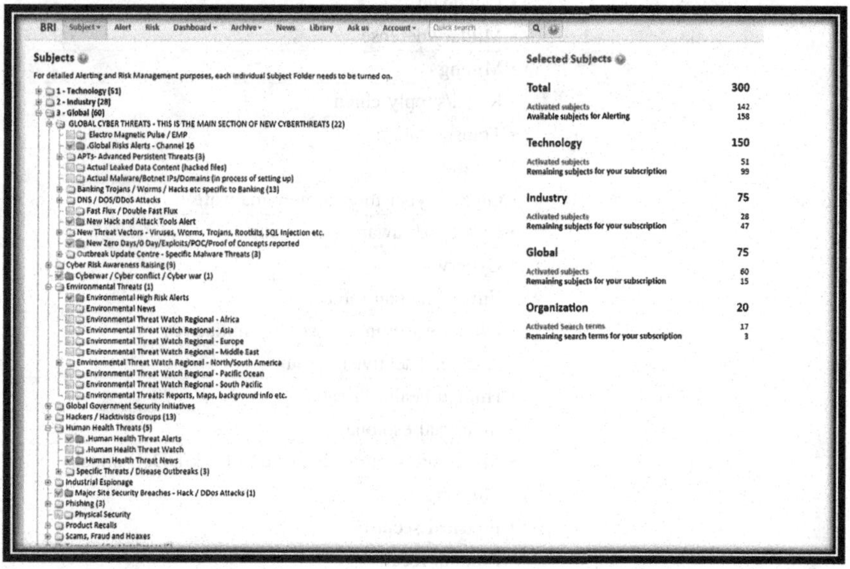

Fig. 1 BRICA subject folders

Last, BRICA includes as a native feature a system for the Online risk Alerting & Management Portal, totally integrated with the Web interface available to its own subscribers, from where the risk intelligence is managed; as per today, the following categories are covered by the different analyst's teams (Table 2).

Giving the nature of this paper, it is pretty obvious how much the attention should be caught by the category "Applications" from the ICT technology risks macro-area.

Table 2 Examples of risk intelligence topics

ICT-technology risks	• Applications
	• Data
	• Network
	• Operating systems
	• Providers
	• Security
	• Systems
Industry-specific intelligence	• Banking/finance
	• Casino's/gaming
	• Chemical
	• Energy
	• Government
	• Health industry
	• ICT/telecom/IXPs/media
	• Industrial control systems (ICS)
	• Law/legal
	• Military/defense
	• Mining
	• Retail/supply chain
	• Transportation
	• Utilities
Global issues	• Global cyber threats, new malware
	• Cyber risk awareness raising
	• Cyberwar
	• Environmental threats
	• Global government security initiatives
	• Hackers/hacktivists groups
	• Human health threats
	• Industrial espionage
	• Major site security breaches—hack/DDoS attacks
	• Phishing
	• Physical security
	• Product recalls
	• Scams, fraud, and hoaxes

Despite speaking about this or any other resources (we have chosen BRICA simply because we use it), we strongly believe that every organization should have a unique data source available, allowing to interact with those different organization's departments, optimizing the effort, the working hours, and the costs of those enterprise's units dedicated to risk analysis and information security, as well as the information flows, helping out in the prevention of IT attacks.

3 Cyber Intelligence (from Closed Sources)

After this overview of open-source-based Cyber Intelligence, let us now talk about the closed sources one. It should be obvious to the reader how much correlation there is between the concept itself of Cyber Intelligence and the issue of insecure programming and bad code: Most of the targets they do realize they have been breached when it is too late, i.e., from a news in the mainstream, a post on Pastebin or an alert from outside the organization itself. That is why there is the need to gain this information, possibly before the incident happens: that is exactly the point and the goal of closed-source-based Cyber Intelligence.

Of course, we are speaking about a totally different approach from the previous one, and typically it is based on annual subscription, which can be resumed as follows:

- AML (Anti Money-Laundering) Intelligence and Consulting: 3C (compromised credit cards), money mules real identities and used bank accounts, etc.
- BOTNETs Intelligence.
- E-CRIME Intelligence.
- MALWARE Intelligence (also "POS" systems).
- THREAT Intelligence.
- TARGETED THREAT Intelligence.

During our activities, we encountered information from the military environment, specifically about credentials which were stolen from compromised computers ("zombies", since they were part of different botnets), with logins and passwords from different sections of Ministries of Defense.

In "slang," this information is called "feeds," and they are coming from targeted attacks, plan, and built specifically for the targeted company; the supplier company, which is much similar to a "private intelligence agency," but operating in the cyber domain, gathers such feeds from different environments, such as the "Cyber" (Cyber Intelligence), Human (HumInt—Human Intelligence, i.e., undercover operations), and Signal (thus, SigInt, Signal Intelligence, i.e., intercepting C&C traffic from botnets, rather than the KU and the C bands for VSAT communication, from satellite data traffic). The sum of all of these feeds brings to the intelligence data, gathered and analyzed by intelligence analysts, which supply a given amount of hours per month to the client's organization.

It is clear that we are speaking about special security services, with a very high level of added value, for which the subscription costs are much higher than the feeds coming from open sources.

Nevertheless, along the years we have witnessed how much the investment approved by different organizations definitely pays back, right at the time in which the *information* allows those organizations to avoid targeted campaigns against themselves, such as hacking campaigns, spear phishing attacks, data breaches, economical fines, class actions, damages to the reputation and image of the organization itself and, last but not least, losing the business continuity.

4 Secure Coding (or Secure Programming)

The previous chapters of this paper helped us to provide a big picture, since it would have been a true mistake to focus only on secure coding, without explaining how much, as of today, the stolen data could *already* be found out of your organization.

Despite all of the hackings which occur non-stop, 24 h a day, during SB +20 years experience with penetration testing and security audits, there are still a very few companies which has applied, i.e., to a secure SLDC (Software Life Development Cycle), believe it or not.

Another clear sign that "something is wrong here" can be realized by the amount of organizations which engage trainers for secure coding trainings. As a mere example, SB provides more than 80 different types of security trainings, and nearly 20 they belong to the secure coding section of our training catalog. Sadly, when we check for figures, only the 2 % of SB customers they buy trainings on secure programming: a very low percentage, which explanation, possibly, may be found in the lack of investment on human resources, and the general approach to outsource the writing of software code to external companies, which often do not care about security issues.

What we have decided to provide over the last section of this paper is our own experience, comparing plus than 1000 penetration testing projects, toward different Web applications, both commercial and open source on.

Still in 2015, we do keep on finding plenty of SQL injections, showing that programmers have not learned the lesson, yet. It is dramatic whenever security team finds this kind of programming mistake—despite the language used, from PHP to. ASP—which allows a remote attacker, without any kind of credentials, to exploit the application's bug and gaining direct access (read, write, delete) sometimes as many as 10.000 tables from the back-end DB.

This leads to the immediate next vulnerability, which can be labeled more a "system administration" one: "Excessive database user grants," allowing attackers

Table 3 Vulnerability table

Vulnerability	Impact	Description
V-001 SQL injection	The attacker can run arbitrary queries on the application DB, and potentially compromise it	User's input is not correctly sanitized and it is inserted into a SQL query
V-002 weak password hashing algorithm	The attacker which has already obtained the password's hash values could obtain clear-text passwords in a very quick timing, using commonly available techniques	Passwords are stored with easy-to-decrypt algorithms and obsolete ones
V-003 weak Oracle passwords	Oracle users are activated with weak passwords, same as the username, or with default values	Passwords of different Oracle users have not been changed, rather are obvious and easy-to-guess, or easy to crack (decrypt)
V-004 excessive database user grants	In the scenario, an attacker is able to execute commands on the DB through the Web application, and the user's privileges of the user assigned to the application are (too) high, allowing the attacker to run major damages from his actions	The user that uses the DB has too high privileges, too much ahead of the standard instructions SELECT, INSERT, DELETE, and UPDATE, which are needed by the application itself in order to work
V-005 multiple cross site scripting	The attacker can steal a legitimate Web application user's session, rather than modify the behavior of the application itself	User input which contains special char sets is not correctly sanitized, thus written in clear text into the application's pages
V-006 missing HTTP only cookie protection	In a XSS scenario, the attacker can get access to the session cookie	The cookie does not have the flag HTTP Only, which avoids access to the cookie via JavaScript
V-007 missing secure cookie protection	In a Man in the Middle and protocol downgrade attack scenario, the session cookie can be intercepted in clear text by the attacker	The Secure flag of the cookie, which prevents it from passing through not encrypted channels, is not enabled
V-008 unsafe sensitive page caching	An attacker can get access to private Web pages, which are saved in the victim's browser cache, rather than in a Proxy through which the requests have passed by	Headers telling to the browser and the proxy to not save passwords in the cache are not specified
V-009 autocomplete not disabled on password field	Passwords from the "password" fields in the application can be saved in clear text on the victim's machines, thus allowing attackers or malwares to obtain them very easily	Wherever the attribute "autocomplete" is not present, rather that it is not set as "off," passwords can be saved by the browser in clear text, allowing autocomplete

(continued)

Table 3 (continued)

Vulnerability	Impact	Description
V-010 missing X-content-type-options header protection	While adding malicious code inside not-malicious files, some browsers could execute it automatically, processing the file in a different manner from the expected one	Not setting the header "Missing X-Content-Type-Options" to "nosniff", some browsers could process malicious code without checking the real file format
V-011 missing X-frame-options header protection	The page can be used by an attacker in order to launch click jacking-type attacks	Without the right headers, the page can be inserted into IFRAMEs in domains which are different from the legal, authorized one
V-012 OPTIONS method enabled	An attacker can learn all of the http methods available in order to plan an attack	The Web Server command OPTIONS returns all of the HTTP commands which can be used
V-013 TRACE method enabled	An attacker may use this method on different attack types, such as bypassing the HTTP Only flag, rather than as a support of different renegotiation vulnerabilities	The debug method TRACE allows to return arbitrary text from the Web Server to the client
V-014 cross-user interaction	A platform's user can get access and modify those data belonging to different users. The same issues apply as well among profiles of different user's type	In some cases, the data visualization and modification pages, related to different types of users, do not verify correctly if the user which is changing them is the owner as well
V-015 SSL/TLS renegotiation	Whenever a Man in the Middle attack happens, the attacker can renegotiate the protocol, thus obtaining data which can be decrypted. In some cases, it is possible also to intercept in clear text the communication	Protocol's renegotiation allows switching the communication from a secure protocol to a weaker one; it is also possible to run reflection-type attacks
V-016 weak SSL/TLS ciphers	An attacker which has obtained the traffic dump can easily decrypt it through widely deployed and available techniques	During the secure SSL/TLS connection, it can be selected ciphers easy to decrypt
V-017 slow Loris DOS vulnerability	A malicious user can overload the Web Server with requests, thus limiting its own availability	The Web Server accepts very long connections and not-existent headers, allowing a malicious user to allocate plenty of resources to the server, for extremely long timings

(continued)

Table 3 (continued)

Vulnerability	Impact	Description
V-018 HTTP session not expire	A malicious user can overload the Web Server with requests, thus limiting its own availability	The Web Server accepts very long connections and not-existent headers, allowing a malicious user to allocate plenty of resources to the server, for extremely long timings
V-019 cross site request forgery	An attacker is able to open a link to the victim, through a validated session, thus executing unexpected, unwanted operations in the application itself	The application does not implement in its forms any anti-CSRF tokens, so that the Web Server cannot verify if the requests are legitimate ones, rather than pushed by an attacker
V-020 potential denial of service using Mail service	A malicious user can send out huge amounts of e-mails to an arbitrary recipient, through the server which is hosting the application	There is no check of the amount of e-mails sent from the automated-sending systems which check for the account confirmation
V-021 change domain on change password mail	An attacker can insert links which are different from those of the organization Web site into the confirmation e-mails, which are addressed to the victim, thus allowing phishing attacks	The application allows the arbitrary choice of the domain which generates the confirmation link for the e-mail matched to an account
V-022 improper error handling	The application shows in clear text the errors' stacktrace, allowing the attacker to understand how it works	Errors' stacktraces are active if some parameters are missing rather than anomalies
V-023 user disclosure	It is possible to run a bruteforce attack on the application's internal IDs (i.e., employee or student's personal identification number), in order to discover those existing ones through the login page	Error messages such as "User does not exist" and "Password incorrect" can be displayed in a different return message from a not-authenticated user
V-024 improper CAPTCHA implementation	The registration CAPTCHA is not deployed correctly, allowing a remote attacker to overload the user's DB, bypassing the anti-automation controls	CAPTCHA is easy to bypass, since it is not deployed correctly
V-025 concurrent session access	Whenever an attacker is connected with the same account of the victim, the authorized user cannot realize, and two concurrent sessions do exist	For some application roles, there is no check if any concurrent sessions for authenticated users do exist
V-026 information disclosure	An attacker can learn useful information related to the Web Server software and its version/release	The Web Server returns in its answers all of the information related to the name and version of the software

to (remotely) obtain hundreds or thousands of user's password; this vulnerability comes very often along with "Weak password hashing algorithm," allowing the attacker to crack user's passwords even faster.

More than often, we then identify vulnerabilities which can be found only when providing the so-called privileged testing approach: right after the typical "black-box" penetration testing approach, in which the team is not provided with any kind of UserIDs/Passwords, we do ask our customers to provide us with 3 different ones, for each kind of "profile," they run on their application, such as "User," "Special User," and "Administrator." The vulnerability is named "Cross-User Interaction" and we keep on finding it nearly every time we run a penetration test, probably because standard security testing companies do not provide their clients with this kind of testing approach.

We can then talk about DoS attacks, meant as bugs in the applications themselves, which allow, i.e., the "Slow Loris DoS" [4] vulnerability, rather than "HTTP Session does not expire" vulnerability: in such cases, SMEs (small and medium size enterprise) are particularly hit, since most of them run their Internet services through standard ADSL lines.

Since the field experiences and the different client's scenarios cannot be resumed so easily, we decided to provide the reader with a table, which highlights most of our standard findings, from those previous 1.000 security testing projects we have carried on along the years.

For each vulnerability (most of them are present in the actual and previous OWASP Top-Ten list), we detail (in the following "Table 3") the vulnerability itself, its impact toward the target organization, and the description.

5   Conclusions

This paper focused on different mistakes which are made by organizations from all over the world, no matter their size and maturity into information security.

While different approaches and the use of different solutions clearly depend on available resources (budgets, amount of IT staff, seniority, and skills of system administrators and programmers), we think that everything should start from tasks which are very easy to deploy inside all of the organizations: internal awareness, and training toward your employees and colleagues.

It is a process which helps out IT and InfoSec people raising the understanding of their everyday activities toward the management, and those colleagues are from different departments, in order to pursue altogether a shared goal: the information security of your organization. A goal that should be an absolute priority for organizations focused on critical security sectors.

References

1. "World's biggest data breach": http://www.informationisbeautiful.net/visualizations/worlds-biggest-data-breaches-hacks/
2. Australian Signals Directorate "Strategies to Mitigate Targeted Cyber Intrusions": http://www.asd.gov.au/infosec/top-mitigations/mitigations-2014-table.htm
3. Cyber open source Intelligence portal: https://brica.de/
4. Slow loris DoS: http://it.wikipedia.org/wiki/Slowloris

Self-validating Bundles for Flexible Data Access Control

P. Gallo, A. Andò and G. Garbo

Abstract Modern cloud-based services offer free or low-cost content sharing with significant advantages for the users but also new issues in privacy and security. To protect sensitive contents (i.e., copyrighted, top secret, and personal data) from the unauthorized access, sophisticated access management systems or/and decryption schemes have been proposed, generally based on trusted applications at client side. These applications work also as access controllers, verifying specific permissions and restrictions accessing user's resources. We propose secure bundles (S-bundles), which encapsulate a behavioral model (provided as bytecode) to define versatile stand-alone access controllers and encoding/decoding/signature schemes. S-bundles contain also ciphered contents, data access policies, and associated metadata. Unlike current solutions, our approach decouples the access policies from the applications installed in the user's platform. S-bundles are multi-platform, by means of trusted bytecode executors. They offer data protection in case of storage in untrusted or honest-but-curious cloud providers.

1 Introduction

Modern ICT frameworks offer users several internet-based computing resources accordingly to the cloud computing paradigm. Cloud applications include e-mail services, data storage, data sharing, and social networking and are consumed by heterogeneous classes of users, through different kinds of devices. On the one hand, cloud computing services deliver several advantages; on the other hand, they introduce new challenges especially on privacy preserving and data security.

P. Gallo (✉) · A. Andò · G. Garbo
DEIM, University of Palermo, Viale delle Scienze Ed. 9, 90124 Palermo, Italy
e-mail: pierluigi.gallo@unipa.it

A. Andò
e-mail: andrea.ando@unipa.it

© Springer International Publishing Switzerland 2016
P. Ciancarini et al. (eds.), *Proceedings of 4th International Conference in Software Engineering for Defence Applications*, Advances in Intelligent Systems and Computing 422, DOI 10.1007/978-3-319-27896-4_23

In fact, most of cloud applications are based on outsourced services and rely on cloud storage. This generally involves cloud providers, beyond the user domain, which have to be considered untrusted, generally honest-but-curious [1, 2]. Cloud services include also access control making data accessible only to authorized users. Several methods have been developed to perform access control managements [3], enforced by the user's machine, or by the Cloud. They permit to define specific permissions and restrictions to the user that allow a fine-grained access control to resources. On the one hand, access control methods that are designed for in-house systems are not suitable for cloud computing applications (as above-mentioned, users, and cloud servers are in different trusted domains [1]); on the other hand, cloud-based access control methods rely on trusted cloud providers. Crypto schemes are used to keep the outsourced sensitive data confidential. However, data will not be available, even to the legitimate user, when he tries to access them from a public device that does not support the specific crypto scheme.

Another privacy issue, related to the data sharing services, is the unauthorized diffusion of sensitive data. When a sensitive content is shared with a limit number of users and, thus, downloaded by one of them (together the decryption key), it can be potentially diffused in a plain form and also consulted in the case of authorization revocation. Digital rights management (DRM) [4] systems employ mechanisms that allow preventing the unauthorized diffusion of contents. To get this target, they usually involve proprietary software [5–7] and/or hardware (Apple iPod, Amazon Kindle and Fire). However, cloud services are offered to several classes of devices, including personal computers, smartphones, tablets, and smart TVs. Therefore, in order to deploy the above-mentioned DRM mechanisms in different devices, a proper software has to be developed for each platform, with limitations about interoperability and portability. DRM systems also employ in some cases well-known crypto scheme and access control management, which require a performing Internet connection.

Sharing information implies data mobility from users' devices to the cloud, forth and back. Furthermore, multiple copies of these data can be duplicated on multiple devices. Unlike existing access control mechanisms, our solution does not depend on crypto/access control capabilities implemented on the device or on the cloud platform. Our novel methodology associates cryptographic methods and access control rules to data. Our access policies move with data and work also on untrusted devices without specific hw/sw prerequisites. The only requirement is about a multi-purpose execution environment [such as a Java virtual machine (JVM)], which is currently available for a large set of devices.

We therefore propose to enclose data into intelligent security boxes, named bundles. These are indivisible entities that contain data, metadata, access rules, and access logic. Using bundles, the access logic and metadata move with data as a second skin and therefore are available no matter the used device. Multiple copies of the data will include all the relative metadata and access policy.

In our typical scheme, the sensitive data file is encapsulated, in a ciphered form, into a unique bundle that is also self-validating (S-bundle). Apart from ciphered data, S-bundle contains the access policies and the needed metadata. In cloud storage environments, the S-bundle can be stored in the outsource provider in place of a simple ciphered data file. Thus, when a user downloads a shared file, an S-bundle, that contains the requested content, is provided. Once the user executes it, S-bundle is able to verify the user credentials (that can be composed of a set of keys) and, in the case of positive verification, to allow user to access to data.

With this scheme, the data stored in cloud servers are continuously protected from the unauthorized access. In fact, the sensitive contents are encapsulated into the S-bundle that uses a logical core to perform the access management. Moreover, the data confidentiality is assured as the data owner stores his/her sensitive data into the S-bundle in an encrypted form. Thanks to the S-bundle, data remain protected even if they have been downloaded and stored in untrusted user platforms. The S-bundle in fact continues to provide the access control to data whatever is the environment in which it is executed.

The S-bundle solution fits the needs of cloud-based services and of digital rights management (DRM) systems, which are used to hinder piracy and the uncontrolled diffusion of copyrighted contents. As a side effect, S-bundles also guarantee the "right to be forgotten" [8, 9], making personal sensitive data no longer available when they are no longer needed for legitimate purposes [10].

The paper is organized as follows: Sect. 2 presents related works, while Sect. 3 introduces the self-validation approach, providing details about the S-bundle workflow. Section 4 provides a description of the S-bundle architecture, and the procedure to build an S-bundle (S-bundle workflow) is described. Section 5 describes an implementation of self-validation system. Finally, Sect. 6 draws conclusions and future work.

2 Related Works

2.1 Security in the Cloud

Attribute-based encryption (ABE) is a modern approach to guarantee privacy, security, and access control in new ICT frameworks [11–14].

ABE schemes are new types of identity-based encryption (IBE) [11, 15] that use private keys whose capability to decipher the ciphertext depends on attributes assigned to users. New classes of ABE consider also data access policies, which can be included into the user private key [12] or into the ciphertext [13]. In detail, in these ABE schemes, the ciphertext and user private keys are issued considering a set of attributes and access policies. Users can decrypt a ciphertext with their key only whether their attributes satisfy the set access policies. With this approach, data are stored in ciphered form into the cloud providers and are accessed only by authorized users with a limited number of distributed keys.

ABE scheme based on ciphertext-policy (CP-ABE) [13] thus partially fulfills the features offered by our proposed approach in terms of access control to data but protects them until they remain in a ciphered form. Once sensitive data are deciphered, also unauthorized users can access them.

2.2 DRM Systems

DRM systems guarantee compliance with the digital license in terms of access and usage rules. DRM prevents the diffusion of the unprotected versions of digital contents. For this reason, DRM systems are used in several commercial scenarios for selling, lending, renting, and streaming of copyrighted contents of different types, such as e-book, games, and multimedia. The DRM world is fragmented because of several standards (DTCP, MPEG-21, OMA, Marlin, etc.) [16–19] and property systems (Adobe Content Server, Apple Fairplay, Microsoft DRM, etc.) [21–24]. This has led to several ecosystems as previously mentioned, and this limits system interoperability and portability.

2.3 Right to Be Forgotten

Internet users have to preserve their privacy that can be compromised by the uncontrolled dissemination of personal information, e-mails, pictures, and videos. Controlling data access and maintaining a complete awareness about them are very difficult, once data are made available on the Web.

A technique to provide the right to be forgotten to user was proposed by Geambasu et al. with a system called Vanish [25]. It exploits the global scale peer-to-peer infrastructure and distributed hash tables (DHT) to map the nodes. Briefly, Vanish encrypts the user's data with a key not known by the user and then destroys it locally. Thus, using a Shamir's secret sharing approach [26], Vanish produces several pieces of the key and sprinkles them at random to nodes of the peer-to-peer networks. DHTs map the nodes that have a piece of key and change continuously. The DHTs are stored only in a limited set of nodes.

After a certain time period, several pieces of the key will naturally disappear because DHT nodes churn or cleanse themselves. When the number of loosed pieces is greater than the complementary (n-k pieces) of the needed minimum number pieces (k), the key cannot be recovered. In this case, user's data cannot be no longer decrypted, and therefore, they become unavailable.

Unfortunately, Vanish has some vulnerabilities, faced by other systems, which are more robust against malicious attacks, such as SafeVanish [27] and Sedas [28].

Our S-bundles natively support time-based access control by adding an expired date that has to be checked against a trusted network time protocol (NTP) server. In the case of unsatisfied time conditions, the S-bundle will not permit the access to the nested data.

3 Self-validation Approach

S-bundles use a self-validation approach: Data are able to self-protect themselves from unauthorized access and become inaccessible beyond a specific expired date. A sophisticated protection wrapper for the data includes several components.

In our vision, S-bundles offer a secure and versatile stand-alone runnable access controller to data. They are able to guarantee protection of sensitive data when they are stored into external server providers or unsecure user environments. S-bundles enclose user's data, crypto schemes, access policies, and information related to authorization mechanisms. Despite the fact that the concept behind S-bundles is technology agnostic, they can be easily implemented using the Java technology. In facts, the JVM is the generic executor, and the bundles themselves are like jar files, containing both logic (the bytecode), user data and metadata (the Manifest file).

S-bundles are intrinsically multi-platform because they rely on a standardized executor (in case of JVM, it is likely already installed in the user platform) to be executed.

Moreover, the content of the S-bundle (data, metadata, and logic) is digitally signed, and the details of the digital signature and digital certificates are also stored inside the bundle.

Figure 1 describes the steps that an authorized user has to carry out to consume the desired contents in the case of the use of S-bundles.

The user requests the desired content to the server provider that answers with the wrapping S-bundle. Once that user has the S-bundle, the data access logic is executed by the virtual machine executor and data are decrypted.

Tamper resistance mechanisms should be adopted in order to certify that the virtual machine executor is trusted and the S-bundle has not been tampered. In order to validate the virtual machine executor, mechanisms derived from trusted computing solutions (untampered hardware and remote attestation) can be employed [29–31]. Other solutions should be oriented on dedicated chips and/or smart cards for the virtual machine executor [32, 33]. The integrity of the S-bundle can be instead evaluated through digital signature verification.

Therefore, once that the S-bundle is executed, first it will verify that the virtual machine is trusted and its integrity is not compromise.

Afterward (in the case of positive validation), S-bundle will ask for user credentials (or other access policies) and these are checked against S-bundle metadata. The credentials can be composed of a proper set of keys, certificates, passwords, etc.

Fig. 1 S-bundle consuming
flowchart

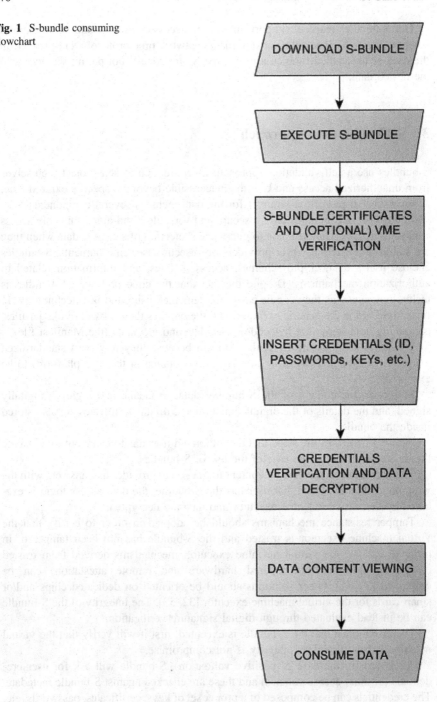

The user can be obtained needed keys through different exchange channels (i.e., e-mails, sms, etc.), needed passwords through specific tokens (one-time password), and needed certificates by proper cards.

Whether the user credentials are valid, the S-bundle grants access to data.

4   S-Bundle Architecture

The self-validating bundle represents a protection wrapper in which user data, metadata, certificates, and a logical core are encapsulated. The S-bundle architecture is depicted in Fig. 2, showing four different components: a data folder, a metadata and information (meta-INF) folder, a certificates folder, and a logical part.

The data folder contains different types of user data (personal, copyrighted, etc.) and with different formats and extensions (.doc, .pdf, .jpeg, .avi, etc.). The flexibility offered by the S-bundle allows to manage different class of files and to define a proper protection to them according to their intended use. As you can note in the scheme of Fig. 2, the data folder is represented in a ciphered form. The encryption of the data folder is essential to guarantee the protection of the user data when the S-bundle is stored in an honest-but-curious (for instance, cloud storage providers) or untrusted (for instance, in the case of unauthorized diffusion) platforms. In fact, in this way, the data contained into the S-bundle are accessible only with the proper credentials. The decryption of the ciphered data folder is performed by the same S-bundle in the case of an authorized access request. The used crypto schemes that can be employed to cipher the data folder are not specified, as they are considered flexible.

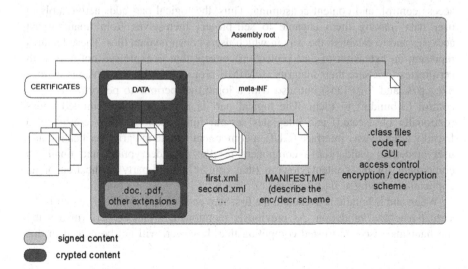

Fig. 2 S-bundle architecture

The data owner is free to choose the crypto scheme that he desires before to generate the S-bundle. Of course, the choice of the crypto scheme involves the contents of other parts of the S-bundle, in particular the meta-INF and the logical part that have to contain the eventual crypto scheme metadata and decryption algorithm, respectively.

The meta-INF folder is contained in the S-bundle in plain form and presents the needed metadata to perform all the S-bundle functionalities. In particular, a file (e.g., the MANIFEST.MF) of the meta-INF folder is dedicated to provide information about the crypto scheme, the version of the S-bundle, the minimum version of the virtual machine executor, and other useful metadata.

In other files (indicated in the scheme of Fig. 2 with *first.xml*, *second.xml*, etc.), information about the requested data and the applications that can open them are reported. In detail, they contain descriptions on the user data format, the enabled third-party applications, and the attributes of the user data. In particular, this attributes offer details on the operations that an application can carry out on the data files. For example, if the attributes grant only file reading, enabled applications will make available only this feature to the user (functionalities of file writing and/or file modifying will be inhibited). However, in order to assure the fair behavior of a third-party application, the latter should be trusted. If an untrusted or malicious application is run by the S-bundle for a data file viewing, users may have more features than how it should have available. In order to overcome this limit and thus to assure the data protection during the content consuming, S-bundles can integrate a trusted data viewer. In this case, functionalities offered by the viewer will follow set data attributes and users will be able to exploit only the proper available features.

The logical part is the core of the S-bundle. In fact, it represents the S-bundle component that allows to accomplish features related to integrity verification, access control, and content consuming. Thus, the logical part adds active tools to user data making them capable to self-protect themselves from unauthorized accesses and to establish the way in which users consume data files. These features represent an extension of data protection concept that is expressed not only with certificates to assure their integrity (S/MIME protocol with the extensions .p7m, . p7c, .p7s, and .p10 [34]) but also with a logic that permits to perform an active control (S-bundle) to them. The logical part is made up with compiled codes (optionally obfuscated) and integrates several functionalities. They consist in a friendly graphic user interface (GUI), a component for the integrity verification and user credential validation, a component for the data decryption, and finally, a component for the content viewing (that can be a third-party application or an application integrated into the bundle).

When and S-bundle is executed, the first step performed by the logical part is the virtual machine validation. As previously mentioned, it can be performed using mechanisms based on trusted computing that, however, will not be treated in this work. Moreover, the S-bundle carries out the integrity certification. This verifies the digital signatures and certificates, contained in the S-bundle. Whether the virtual machine executor is not trusted or S-bundle integrity is not confirmed, the logical part advises the user about inconsistencies and thus stops the S-bundle. Conversely, the GUI is launched.

GUI is the first interface that users encounter after the execution of the S-bundle. It has the task of requiring the user credentials. The user will give these information by inserting text into proper text fields (such as IDs, passwords, one-time password, etc.) or by providing the path for files that contain required certificates, keys, and/or other identification data. A verification tool of the S-bundle will thus examine the inserted credentials. In particular, it will verify the identity, the access policies, and the keys (inserted by the user). In particular, the keys will be used to perform the decryption process. The mechanisms for the access control procedures depend on sharing typology, used crypto scheme, access policy details, and key distribution methods. When data have to be shared with a single user, building the S-bundle is very convenient and an asymmetric crypto scheme can be considered. In this case, the data will be encrypted with the public key of the end user. In fact, through the private end user key, the S-bundle will be able to identify the user to decipher the data and to provide them in a plain form according to the set access policies (that can consider a limited number of access and an expired date).

In the other hand, when the data sharing is oriented to more than one user, mechanism such as one-time password, online user identification, or other suitable solutions should be considered for the S-bundle. For instance, a symmetric cryptography can be used in these cases. However, a distribution key management has to be considered. Moreover, for the user identification, other mechanisms should be employed such as an online user authentication (e.g., carried out by an external server with secure protocols).

Novel encryption techniques introduced in Sect. 2.1 and called attribute-based encryption (ABE) [11–14] can be fruitfully used for access control and decryption mechanisms of the S-bundle. In particular, there are ABE classes that are capable to provide a fine-grained access control to resources. We refer to the key-policy ABE (KP-ABE) [12] and to the ciphertext-policy ABE (CP-ABE) [13]. In these ABE classes, the ciphertext and user private keys are issued considering a set of attributes and an access policy. So that a user can decrypt a ciphertext with his/her key only whether his/her attributes satisfy the set access policy. The attributes and the access policy can be contained into the ciphertext and the private keys, respectively (KP-ABE), or into private keys and ciphertext, respectively (CP-ABE).

The S-bundle flexibility permits to use several alternatives for the access control and encryption schemes, without any pre-defined choice.

The encryption component contains computer algorithms assolving data decryption as well as data and identity control. Generally, its complexity is related to crypto scheme. The output of the data decryption component is data in plain form. The latter will be open to viewer tools allowing the user to consume it. As previously mentioned, the viewer tool can be external to the S-bundle or can be integrated in it. Attributes or licenses of use permit to define how the user can consume contents. Of course, in the case of viewer tool integrated into the S-bundle, the rightful use of contents from the user is easier to assure. However, integrating a viewer into the S-bundle can increase a lot its size (this depends on data files format). A third-party application excludes this drawback but introduces a security issue related to its trusted behavior. Nevertheless, with a tool validation step (that may be carry out by the S-bundle), also in the case of third-party application, a fair use of data can be obtained. In order to get the tool validation, the S-bundle has to perform a verification procedure to certificate that the version of the third-party tool is trusted. This step can be performed with the aid of an external secure environment consulted by S-bundle, by means of Internet connection, and by using secure protocols [29].

4.1 S-Bundle Workflow

In Sect. 3, we have introduced details on steps that a user should carry out in order to be able to consume a content in a self-validation scenario. We have also described that the main element of this approach is the S-bundle. In Sect. 2, we have described how the S-bundle is composed and what features makes available. According to the specifications introduced in Sect. 3, we can also present needed steps to generate an S-bundle, starting from data you want share. Figure 3 depicts the flowchart that describes how an S-bundle is build.

A user that wants to encapsulate data into the S-bundle is able to exploit a wide range of configuration settings. In fact, before the building of the final version of the S-bundle (with the signature), the user can select several settings: the crypto scheme, who can access to data (i.e., access policy), how contents have to consume (license of use), and what viewer can be used to open data (that can be also integrated into the bundle).

A proper S-bundle builder (S-builder) is provided to the user to aid him/her during the generation steps. Moreover, S-builder is capable to perform other functionalities such as the data encryption and (optionally) the key distribution. In particular, after a user defines the crypto scheme and the access policies, the S-builder will perform data encryption. Once data are encrypted, the user will be able to select the eventual key management. The same S-builder will distribute keys to the end users only after that the S-bundle is signed.

During the procedure to generate the S-bundle, a user can also define what third-party software will be enabled to open user data. Alternatively, a user can also select to integrate a proper viewer into the S-bundle.

Fig. 3 S-bundle building
flowchart

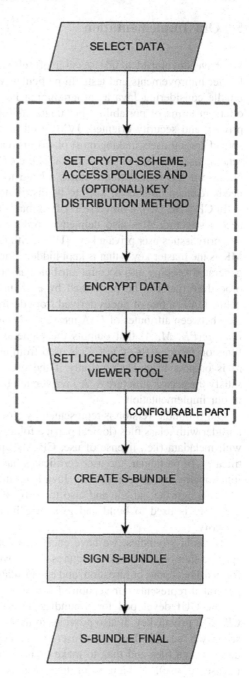

5   Our Implementation

We have developed a first version of self-validation system that will be under further improvements and tests. In particular, we have designed our self-validating bundle considering Java as programming language, for the functionalities that it offers in terms of portability and upgradability and for several available libraries privacy and security oriented. JVM, used to execute Java codes, is further widespread among users making most platforms ready to execute S-bundles. The crypto scheme used for our implementation is the CP-ABE (presented in Sect. 2.1). This because it has remarkable features related to the fine-grained access control and needs reduced number of keys to be distributed.

In CP-ABE [13], a universe U of attributes is considered. Moreover, from U, an access structure A is also defined in order to set the desired access policy. An authority issues user private keys (UKs) with the function $Setup(PK, MK, U)$ where MK is the master key—that is kept hidden—and PK is the public key. Every UK is generated keeping into account attributes associated to the user. For the encryption procedure that can be performed by each user, an access structure A has to be defined. A is a tree of nodes derived from desired logic relations (AND, OR, 2 of 3, etc.) between attributes of U. A message M can be thus encrypted with the function $Encrypt(PK, M, A)$ that outputs the ciphertext CT. In order to decrypt CT, the function $Decrypt(PK, CT, UK)$ is used from users. In particular, the plain message M is outputs from $Decrypt$ only if the user attributes (considered into the UK) satisfy the access structure A. A Java version of previous CP_ABE functions is used in our implementation.

Our S-bundle version is represented by a runnable .jar file in which is contained: a folder with .class files (logical part); a folder with ciphered user data, and a folder with metadata (i.e., license of use, CP-ABE public key, etc.), signatures and certificates. In particular, the user certificates have been generated with the "keytool" functionality, made available by Java Runtime Environment. In the same way, in order to sign the .jar file and also to verify it, the "jarsigner" tool has been used. This tool is used to build and to sign S-bundles and is also to verify S-bundle integrity.

To build s-bundles, we have encrypted data with the CP-ABE scheme with a specific number of access structures. We have also set the needed metadata, performed Java codes obfuscation, and encapsulated and signed S-bundles. The .jar file generated represents our version of S-bundle.

The GUI developed for S-bundles allows users to insert the path of his/her CP-ABE private key. It also provides to users eventual advises derived from errors, security issues, etc. Having the user private key, the S-bundle logical part is able to decrypt data files and then to make them available to user (with limits defined in license of use). A viewer for text files is integrated into S-bundles, whereas third-party applications (already installed in the user platform) are used for the other classes of files. The S-bundle logical part protects the open files to prevent the user to store a plain version of them.

The JVM validation is another important procedure to accomplish in order to avoid possible attempts to compromise S-bundles. Nonetheless, we have chosen not to develop this module for this first implementation in which JVM is considered trusted. As JVM, third-party applications are considered trusted. Ad hoc mechanisms for JVM and third-party applications validation to be included into S-bundles will be investigated and developed in future works. However, the lack of these two modules does not compromise the concept of self-validation approach proposed in this paper.

6 Conclusions

We have presented an access controller-based self-validation approach. The data are encapsulated in a wrapper (S-bundle), that is stand-alone, multi-platform, digitally signed, uses crypto schemes to cipher the data (selected by the user), and optionally equipped with a tamper protection. It offers a logical part able to perform the access management, the data decryption, and the managing of data consumption. Thanks to the above-mentioned logical part, data are able to self-protect themselves from unauthorized accesses. A user that wants to access to data has to execute the S-bundle that with the logical part is able to verify the user credentials (id, password, keys, etc.), to decrypt the data (in the case of positive verification), and also to provide the data contents to the user according to the set license of use. Data can be opened with a third-party application or with a viewer integrated into the S-bundle. Software verification procedures are kept into account to validate the virtual machine executor and third-party applications in order to assure their trusted behavior.

Our approach represents a flexible and secure alternative to the current commercial solutions in every environment in which the security and privacy of the data is a main topic. Cloud computing and data sharing, electronic health, and Armed Force are just some examples of application scenarios.

Acknowledgment This work has been partially founded by the national project PON04a2_C SMART HEALTH—CLUSTER OSDH—SMART FSE—STAYWELL, which applies these results to several e-health scenarios.

References

1. Li J, Chen X, Li J, Jia C, Ma J, Lou W (2013) Fine-grained access control system based on outsourced attribute-based encryption. In: Proceedings of the 8th European symposium on research in computer security (ESORICS 2013), Egham, UK, Computer Security—ESORICS 2013, Lecture Notes in Computer Science, Springer, vol 8134, pp 592–609
2. Zhu Y, Ma D, Hu CJ, Huang D (2013) How to use attribute-based encryption to implement role-based access control in the cloud. In: Proceedings of the international workshop on security in cloud computing 2013 (cloud computing '13), ACM, pp 33–40

3. McDaniel P, Prakash A (2006) Methods and limitations of security policy reconciliation. ACM Trans Inf Syst Secur 9(3):259–291
4. Liu Q, Safavi-Naini R, Sheppard NP (2003) Digital rights management for content distribution. In: Proceedings of the Australasian information security workshop on ACSW frontiers, Feb 2003, Adelaide (Australia), vol 21, pp 49–58
5. www.adobe.com/solutions/ebook/content-server.html
6. "FairPlay", www.wikipedia.com
7. How FairPlay Works (2007) Apple's iTunes DRM dilemma. www.roughlydrafted.com
8. Weber RH (2011) The right to be forgotten—more than a pandora's box? J Intellect Property Inf Technol Electron Commer Law 2(2)
9. European Commission, Factsheet on the 'right to be forgotten' ruling, (C-131/12)
10. European Commission (2010) Communication from the commission to the European parliament, the council, the economic and social committee and the committee of the regions, COM(2010) 609 final, Brussels, 2010
11. Sahai A, Waters B (2005) Fuzzy identity based encryption. Adv Cryptol V EUROCRYPT 3494:457–473
12. Goyal V, Pandey O, Sahai A, Waters B (2006) Attribute-based encryption for fine-grained access control of encrypted data. In: Proceedings of the 13th ACM conference on computer and communications security, pp 89–98
13. Bethencourt J, Sahai A, Waters B (2007) Ciphertext-policy attribute-based encryption. In: Proceedings of IEEE symposium on security and privacy, pp 321V334
14. Wang G, Liu Q, Wu J (1985) Hierarchical attribute-based encryption for fine-grained access control in cloud storage services. In: Proceedings of the 17th ACM conference on computer and communications security, pp 735–737, 2010
15. Shamir A (1985) Identity-based cryptosystems and signature schemes. In: Proceedings of CRYPTO 84, Lecture Notes in Computer Science, Springer, Berlin, vol 196, pp 47–53
16. www.dtcp.com
17. mpeg.chiariglione.org/standards/mpeg-21/mpeg-21.html
18. openmobilealliance.org
19. www.marlin-community.com
20. www.adobe.com/solutions/ebook/content-server.html
21. en.wikipedia.org/wiki/Adobe_Content_Server
22. en.wikipedia.org/wiki/FairPlay
23. en.wikipedia.org/wiki/Windows_Media_DRM
24. en.wikipedia.org/wiki/Active_Directory_Rights_Management_Services
25. Geambasu R, Kohno T, Levy AA, Levy HM (2009) Vanish: increasing data privacy with self-destructing data. In: Proceedings of the USENIX security symposium. Montreal, Canada, Aug 2009
26. Shamir A (1979) How to share a secret. Commun ACM 22(11):612–613
27. Zeng L, Shi Z, Xu S, Feng D (2010) SafeVanish: an improved data self-destruction for protecting data privacy. In: Proceeding of the second international conference on cloud computing technology and science (CloudCom). IEEE, pp 521, 528
28. Zeng L, Chen S, Wei Q, Feng D (2013) SeDas: a self-destructing data system based on active storage framework. IEEE Trans Magn 49(6):2548, 2554
29. Garfinkel T, Pfaff B, Chow J, Rosenblum M, Boneh D (2003) Terra: a virtual machine-based platform for trusted computing. In: Proceedings of the nineteenth ACM symposium on operating systems principles. ACM Press, Bolton Landing, NY, USA, pp 193–206
30. www.trustedcomputinggroup.org
31. Schellekens D, Wyseur B, Preneel B (2008) Remote attestation on legacy operating systems with trusted platform modules. Sci Comput Program 74(1–2):13–22
32. www.arm.com/products/processors/technologies/jazelle.php
33. www.oracle.com/technetwork/java/embedded/javacard
34. Dusse S, Hoffman P, Ramsdell B, Lundblade L, Repka L (1998) S/MIME version 2 message specification, RFC 2311

Improving Bug Predictions in Multicore Cyber-Physical Systems

Paolo Ciancarini, Francesco Poggi, Davide Rossi and Alberto Sillitti

Abstract As physical limits began to negate the assumption known as Moore's law, chip manufacturers started focusing on multicore architectures as the main solution to improve the processing power in modern computers. Today, multicore CPUs are commonly found in servers, PCs, smartphone, cars, airplanes, and home appliances. As this happens, more and more programs are designed with some degree of parallelism to take advantage of these implicitly concurrent architectures. In this context, new challenges are presented to software engineers. For example, software validation becomes much more expensive (since testing concurrency is difficult) and strategies such as bug prediction could be used to better focus the effort during the development process. However, most of the existing bug prediction approaches have been designed with sequential programs in mind. In this paper, we propose a novel set of concurrency-related source code metrics to be used as the basis for bug prediction methods; we discuss our approach with respect to the existing state of the art, and we outline the research challenges that have to be addressed to realize our goal.

P. Ciancarini (✉) · F. Poggi · D. Rossi
Department of Computer Science and Engineering, University of Bologna, Bologna, Italy
e-mail: paolo.ciancarini@unibo.it

F. Poggi
e-mail: francesco.poggi5@unibo.it

D. Rossi
e-mail: daviderossi@unibo.it

A. Sillitti
Center for Applied Software Engineering and CINI, Bologna, Italy
e-mail: alberto@case-research.it

© Springer International Publishing Switzerland 2016
P. Ciancarini et al. (eds.), *Proceedings of 4th International Conference in Software Engineering for Defence Applications*, Advances in Intelligent Systems and Computing 422, DOI 10.1007/978-3-319-27896-4_24

1   Introduction

Parallel programs have been around for a long time, but they are becoming mainstream only now, with the ubiquitous introduction of multicore CPUs. In the perspective of the software engineer, this opens up new possibilities but also new issues to deal with.

One of the issues we perceive as most critical is related to software validation. Testing is the validation strategy that is adopted in most software development methods to limit bugs-related problems: Extensive testing is known to limit the number of failures affecting deployed software systems [1]. Testing can be performed statically (i.e., analyzing the code, usually by the mean of formal approaches, without executing it) or dynamically (executing the code within a so-called test case). Several research works have focused on static analysis for concurrent programs, and there are tools that implement such research outcomes. However, they are of limited practical applicability due to the specific skills and the extra effort required to the developers (e.g., definition of formal specifications using specific languages such as Z [2]), and most developers prefer to adopt dynamic testing techniques (e.g., unit testing using tools such as these of the xUnit family) as the more pragmatic solution. Dynamic testing is usually applied trying to maximize parameters such as code/branch coverage [3]; the underlying idea is to make sure that a test set is able to execute at least once all the different parts of the program in a way that no code goes untested. Even if this approach is far from being able to find all the possible defects, it is widely adopted and generally accepted in most domains. Parallel programs introduce a new dimension to the coverage concept: Not only all the possible sequences of operations for a piece of code have to be taken into account, but also all the sequences for all the active threads and all their interleavings. This makes it practically impossible to create (and run) test sets with a satisfactory level of coverage for most parallel code [4]. In [4], large software projects exploiting concurrency are deeply analyzed to characterize concurrency-related bugs; the experience accumulated in such a work led the authors to suggest three strategies to address concurrency-related bugs:

1. create bug detection methods that take into account concurrency-related problems;
2. improve testing and model-checking approaches;
3. improve the design of concurrent programming languages.

Given the aforementioned context, we believe too that bug prediction strategies, allowing focusing testing efforts where they are needed the most, will become even more relevant. While in the past few years a solid amount of research has been devoted to bug prediction methods and tools, very little has been done to take into account the specificities introduced by the exploitation of parallelism. We argue that it should be possible to take advantage of concurrency-specific aspects of programs to improve bug detection in parallel programs. More specifically, we propose the following research questions:

- Q1: Can we devise complexity metrics for parallel programs that can be related to the number of concurrency-related bugs?
- Q2: Can bug prediction make use of the specificities of parallel programs to obtain better precision and/or recall?
- Q3: Can bug prediction give indications on concurrency-related bugs (the ones that are more difficult to address using testing)?

This paper is structured as follows: In the next section, we briefly summarize the state of the art in bug prediction and concurrency; in Sect. 2, we focus on the research on bug prediction that takes into account concurrency-specific metrics. Section 3 presents our initial approach to improve existing methods. Finally, Sect. 4 draws the conclusions.

2 State of the Art

2.1 Bug Prediction

The general approach taken by defect prediction methods is based on the identification of correlations between a set of code and/or process metrics and the number of defects.

Approaches based on code metrics usually take into account the measures proposed in the late 1970s by Halstead [5], McCabe [6], and others [7]. Measures can be collected easily through simple programs parsing the code under scrutiny.

Approaches based on process analysis use a wide variety of measures from different kinds of sources including complexity/frequency of code changes collected by code revision tools, history from bug repositories, and many more [8].

Moreover, there are several mixed approaches that include both code and process metrics [9–13].

There are already plenty of works dealing with the analysis and the comparison of the different approaches; therefore, we just focus on the ones we consider most valuable literature reviews available.

A good review on fault prediction studies (focused on software metrics-based approaches) is presented in [14]: The authors show that, after year 2005, most works use machine learning-based correlation approaches (66 %), followed by works using hybrid machine learning/statistical approaches (17 %), and purely statistical approaches (14 %).

They also discuss the nature of the code metrics used with respect to the granularity of the elements analyzed (file, component, class, method, etc.) and argue that more class-related metrics should be used to better align the prediction method with the object-oriented paradigm (when used).

A more recent work is [15] in which the authors argue that bug prediction methods should be compared on the basis of an established benchmark (which they propose). A good introduction to existing approaches (that also include process-based metrics) is provided along with related bibliographical references. A problem with the proposed benchmark is that it uses a code base composed by software written uniquely in the Java programming language. While this allows a more uniform application on the code metrics that are used to compare different prediction methods, it also makes most of the work hardly applicable to software projects using other programming languages.

In both [14, 15], and all the works they reference, no concurrency-related measure is used.

2.2 Bug Prediction and Concurrency

Concurrency introduces a further complexity dimension to software systems; several studies show that concurrency-related bugs can be elusive and difficult to track [4, 16–19], which has an impact on the costs of the development process. Other works not directly linked with predicting bugs, but nevertheless worth citing for the affinity of challenge they face are [4, 20, 21]: In particular, they provide insights and a comprehensive view on bugs taxonomy, testing criteria, and tools in concurrent programs.

As previously stated, there is little published research work regarding concurrency-related bugs prediction. However, there are some interesting studies.

In [16], the authors focus on Mandelbugs [22] in complex software systems. Mandelbugs are defined as faults that are triggered by complex conditions, such as interaction with hardware and other software, and timing or ordering of events (as is usually the case for concurrency-related bugs). A fault prediction model is proposed that, among classic software complexity measures, exploits concurrency-related constructs, such as the number of synchronized methods and blocks in a class (the assumed programming language is Java).

A machine learning-based approach is then used either to classify software modules (in this case "Mandelbug-prone" or "Mandelbug-free") or to rank modules with respect to the (expected) number of Mandelbugs that are present. A validation is then presented, and the result is that the introduction of novel metrics based on concurrency (but also I/O operations and exception handling constructs) allows the creation of fault predictors having higher accuracy with respect to traditional metrics.

Reference [19] focuses explicitly on concurrency bug prediction. The approach presented is mainly process-based: Using several metrics (e.g., number and size of patches, elapsed days, modified files, number of comments, number of developers, and severity), the prediction models are able to predict the number and type of concurrency bugs injected by developers. The types of bugs taken into account include atomicity violations, order violations, data races, and deadlocks. For bug number prediction, two models are developed: one based on generalized linear regression and the other on time series. The models predict the number of concurrency bugs that will appear in future releases.

For bug-type prediction, a machine learning approach, based on classifiers, is used. The method applies standard information retrieval techniques to extract relevant keywords from bug reports.

Three large popular open-source projects (Mozilla, KDE, and Apache) are the dataset taken into consideration.

To our knowledge, there are no works focusing on the development of prediction models based on code metrics.

3 Our Approach

Due to the limited research in the area and the increasing interest due to the technological evolution, the development of novel approaches to reliably predict faults in concurrent code is needed. Our research effort is focused on the development of a set of basic and easy-to-collect code metrics that are able to provide a baseline for the evaluation and the comparison of pieces of code. We propose an initial set of four metrics we think can improve the reliability of prediction models. We have designed such set of metrics with the following characteristics:

- **Language independent**: There is a wide range of languages that are popular to develop concurrent code in different domains (e.g., enterprise and Web applications, embedded systems, and mobile); therefore, supporting the most common languages is of paramount importance. The design of the metrics took into consideration the following languages: C/C++, Java, C#, and Ada. Due to the structure of the designed metrics, we think that most of procedural and object-oriented languages can be supported; we do not have investigated them at present, but we would like to analyze their applicability to further languages in the future.
- **Usable in both procedural and object-oriented code**: Even if object-oriented code is widely adopted, there are domains where procedural languages (e.g., C) are still major players. Therefore, our metrics need to be usable with both paradigms.
- **Easy to extract**: Metrics collection is a time-expensive activity and can be effectively implemented only if it is fully automated. Therefore, it is required that the designed metrics can be easily collected through simple static code analysis techniques with a fast processing. In this way, it is possible to analyze large code bases (including the time evolution through the analysis of entire version control repositories).
- **Process independent**: The proposed metrics do not rely on any process information; therefore, they can be applied even if no structured information on the development process is present. This property is useful for the analysis of legacy code where process information was not stored anywhere or it is not available anymore. However, supplementing our metrics set with process information (e.g., from version control systems) could provide benefits that need to be investigated.

We are aiming at identifying a set of metrics that are able to provide a comprehensive overview of the structure of the code from the concurrency point of view, considering aspects such as the complexity of the code and it size. To do that, we take inspiration from the well-known metrics already available in the literature (e.g., McCabe cyclomatic complexity) and we adapt them to the novel environment. We acknowledge that there are a number of limitations and pitfalls in doing that. Therefore, a deep and careful assessment of such metrics is needed to evaluate their effectiveness in improving defect models. In this paper, we propose this set of metrics that is based on a very preliminary evaluation of their properties and effectiveness.

We propose the following metrics:

- **Forkomatic complexity**: It aims at measuring the synchronization complexity caused by the presence of different threads. Unlike its name, it does not measure the number of forks in the code but the number of synchronization primitives in a function/method. The rationale behind such a measure is related to the fact that in most of the cases, the complexity related to concurrency is related to the synchronization of the different activities performed rather than to the creation of such activities. Therefore, we counted the synchronization primitives as a reasonable measure of the complexity of the concurrent code.

- **Sync number**: It aims at measuring the complexity of the synchronization structures. It is calculated as the number of synchronization pairs (pairs of Ps and Vs in traditional semaphores-inspired concurrency control) at the same annidation level. Unlike the forkomatic complexity, this metric considers the structure of the code identifying the correspondence between the synchronization primitives. Therefore, it provides a measure of the complexity of the structures used to implement the concurrency.

- **Critical complexity**: It aims at measuring the amount of the concurrency complexity in the code. It is calculated as the number of critical sections in a function/method/class. It provides a measure of how many critical sections exist in a specific area of the code helping at identifying how distributed is the concurrency logic in the code.

- **Critical density**: It aims at measuring the impact of the size of the critical sections compared to the overall code. It is measured as the ratio between the LOC (Lines of Code) in the critical sections and the total LOC. It provides a measure of the amount of code that requires concurrency control compared to the regular code.

Our proposal is based on ideas gathered from the existing literature, personal experiences, and common sense. An extensive validation work is required to assess its effectiveness. Specifically, we need to access the ability of the metrics listed above in improving the reliability of defect prediction models in concurrent code. This includes an investigation about the appropriate level of granularity (e.g., method, class) to adopt for each metric to provide a suitable set of guidelines in their usage. One of the problems we are facing is related to the choice of the dataset to use for the validation. Although a publicly available code repository would be desirable, common choices such as the Promise repository [23] have the drawback of containing little or no concurrent code. We are investigating alternatives such as using popular open-source projects as proposed in [17, 19] along with the publicly available code from open-source concurrent software systems such as game-playing engines (e.g., multicore chess programs such as Octochess[1] and Stockfish Chess[2]).

To evaluate the strength and efficacy of the proposed metrics, we plan to arrange/set up a testing framework to easily compare our results with previous work on software metrics and bug predictions. In particular, we plan to compare our indicators with both classic sequential metrics (e.g., Halstead [5], McCabe [6], Chidamber-Kemerer [24], and others [7]) and more recent works on concurrent software metrics, such as [19]. Moreover, the choice of the dataset is of paramount importance for the evaluation task too, since focusing on popular open-source systems that have had a history of more than 10 years (e.g., Mozilla, KDE, Apache) would let us to leverage the information about their evolution contained (and publicly available) versioning systems. The idea is to calculate our metrics at predefined intervals (e.g., every year, or every stable release) and test their efficacy in predicting bugs by analyzing real bugfix processing data (e.g., bug reports, number and distribution of commits, type of modification, number of edited files and changed lines of codes), following an approach similar to our previous work [25] and using a modified version of the tools we developed to collect the data of [26, 27]. Finally, the development of analysis tools based on open-source parsers that work on the abstract representation of the source code (e.g., those provided by the JDT (for Java) or CDT (for C/C++) projects) represents a robust choice for building general solutions that can be easily extended to different languages, as described in [28]. For this purpose, we are also considering other extendable tools for collecting software metrics, generating dependency graphs, and analyzing software evolution, such as Metrix++[3] and Analyzo [29].

4 Conclusions

This paper has proposed a new set of code metrics aiming at improving the reliability of defect prediction models in the case of concurrent code. As in any new proposal of metrics, the development of a baseline from real code is needed. For this reason, an extensive empirical evaluation is in progress to assess the effectiveness of such metrics and build a first set of reference metrics. In particular, in Sect. 1, we have identified three research questions that guide our empirical

---

[1]http://octochess.org.
[2]http://stockfishchess.org.
[3]http://metrixplusplus.sourceforge.net.

evaluation of the proposed metrics. Till now, we do not have enough data to provide an answer to such questions, but our very preliminary results are very promising and make us confident in finding such answers in a short time.

References

1. Bertolino A (2007) Software testing research: achievements, challenges, dreams. In: 2007 future of software engineering. IEEE Computer Society, pp. 85–103
2. Spivey JM (1989) The z notation: a reference manual. Prentice-Hall International Series in Computer Science
3. Zhu H, Hall PAV, May JHR (1997) Software unit test coverage and adequacy. J ACM Comput 29(4):366–427
4. Lu S, Park S, Seo E, Zhou Y (2008) Learning from mistakes: a comprehensive study on real world concurrency bug characteristics. In: ACM Sigplan Notices, vol 43, ACM, pp 329–339
5. Halstead MH (1977) Elements of software science (Operating and Programming Systems Series). Elsevier Science Inc, Amsterdam
6. McCabe TJ (1976) A complexity measure. In: Proceedings of the 2nd international conference on software engineering, ICSE '76. IEEE Computer Society Press, p 407
7. Fenton N, Bieman J (1997) Software metrics: a rigorous and practical approach. CRC Press, Boca Raton, FL
8. Fronza I, Sillitti A, Succi G, Vlasenko J, Terho M (2013) Failure prediction based on log files using random indexing and support vector machines. J Syst Soft Elsevier 86(1):2–11
9. Coman I, Sillitti A, Succi G (2008) Investigating the usefulness of pair-programming in a mature agile team. In: 9th international conference on extreme programming and agile processes in software engineering (XP2008), Limerick, Ireland, 10–14 June 2008
10. Pedrycz W, Succi G, Sillitti A, Iljazi J (2015) Data description: a general framework of information granules. Knowl Based Syst 80:98–108 (Elsevier)
11. Coman I, Robillard PN, Sillitti A, Succi G (2014) Cooperation, collaboration and pair-programming: field studies on back-up behavior. J Syst Soft 91(5):124–134 (Elsevier)
12. Di Bella E, Fronza I, Phaphoom N, Sillitti A, Succi G, Vlasenko J (2013) Pair programming and software defects—a large, industrial case study. IEEE Trans Soft Eng 39(7):930–953
13. Fronza I, Sillitti A, Succi G (2009) An interpretation of the results of the analysis of pair programming during novices integration in a team. In: 3rd international symposium on empirical software engineering and measurement (ESEM 2009), Lake Buena Vista, FL, USA, 15–16 Oct 2009
14. Catal C, Diri B (2009) A systematic review of software fault prediction studies. Exp Syst Appl 36(4):7346–7354
15. D'Ambros M, Lanza M, Robbes R (2010) An extensive comparison of bug prediction approaches. In: 2010 7th IEEE working conference on mining software repositories (MSR), IEEE, pp 31–41
16. Carrozza G, Cotroneo D, Natella R, Pietrantuono R, Russo S (2013) Analysis and prediction of mandelbugs in an industrial software system. In: 2013 IEEE sixth international conference on Software testing, verification and validation (ICST), IEEE, pp 262–271
17. Fonseca P, Li C, Rodrigues R (2011) Finding complex concurrency bugs in large multi-threaded applications. In: Proceedings of the sixth conference on computer systems, ACM, pp 215–228
18. Shihab E, Mockus A, Kamei Y, Adams B, Hassan AE (2011) High-impact defects: a study of breakage and surprise defects. In: Proceedings of the 19th ACM SIGSOFT symposium and the 13th European conference on Foundations of software engineering. ACM, pp 300–310

19. Zhou B, Neamtiu I, Gupta R (2015) Predicting concurrency bugs: how many, what kind and where are they? In: Proceedings of the 19th international conference on evaluation and assessment in software engineering, ACM, p 6
20. Brito M, Felizardo KR, Souza P, Souza S (2010), Concurrent software testing: a systematic review. In: Testing Software and Systems: Short Papers, p 79
21. Souza SR, Brito MA, Silva RA, Souza PS, Zaluska E (2011) Research in concurrent software testing: a systematic review. In: Proceedings of the workshop on parallel and distributed systems: testing, analysis, and debugging, ACM, pp 1–5
22. Grottke M, Trivedi KS (2007) Fighting bugs: remove, retry, replicate, and rejuvenate. IEEE Comp 40(2):107–109
23. Menzies T, Caglayan B, Kocaguneli E, Krall J, Peters F, Turhan B (2012) The promise repository of empirical software engineering data
24. Chidamber SR, Kemerer CF (1994) A metrics suite for object oriented design. IEEE Trans Soft Eng 20(6):476–493
25. Di Bella E, Sillitti A, Succi G (1994) A multivariate classification of open source developers. IEEE Trans Soft Eng 221:72–83
26. Scotto M, Sillitti A, Succi G, Vernazza T (2006) A non-invasive approach to product metrics collection. J Syst Arch 52(11):668–675
27. Scotto M, Sillitti A, Succi G, Vernazza T (2004) A relational approach to software metrics. In: 19th ACM symposium on applied computing (SAC 2004), Nicosia, Cyprus, 14–17 Mar 2004
28. Janes A, Piatov D, Sillitti A, Succi G (2013) How to calculate software metrics for multiple languages using open source parsers. Springer, Berlin, pp 264–270
29. Terceiro A, Costa J, Miranda J, Meirelles P, Rios LR, Almeida L, Chavez C, Kon F (2010) Analizo: an extensible multi-language source code analysis and visualization toolkit. In: Brazilian conference on software: theory and practice (Tools Session)

Predicting the Fate of Requirements in Embedded Domains

Witold Pedrycz, Joana Iljazi, Alberto Sillitti and Giancarlo Succi

Abstract The prediction of the fate of a requirement has a visible impact on the use of the resources associated with it. Starting the implementation of a requirement that does not get inserted in the final product implies that we waste the resources associated with it. Moreover, we also create code and other artifacts that may be difficult to identify and delete once the decision of not to proceed with such requirement has been made, if it will ever been made explicitly instead of simply letting the requirement to fall in the oblivion. Furthermore, an additional requirement creates an increased complexity in managing the system underdevelopment. Therefore, it would be of immense benefit to predict the fate of such requirement as early as possible. Still, such prediction does not seem to be feasible and has not been subject yet to a significant investigation. In this work, we propose an approach to build such a prediction in a domain in which it is of particular benefit—the embedded systems domain, where typically the cost of errors is higher due to its direct impact on hardware. To determine whether a requirement will fail, we consider simply the history of the operations it underwent, treating the requirement as a black box. We use logistic regression to discriminate among the failures. We verify the model on more than 80,000 logs for a development process of over 10 years used in an Italian company operating in an embedded domain. The results are interesting and worth follow-up analysis and extensions to new datasets.

W. Pedrycz · J. Iljazi
Department of Electrical and Computer Engineering, The University of Alberta,
Edmonton, Canada

A. Sillitti (✉)
Center for Applied Software Engineering and CINI, Bolzano, Italy
e-mail: alberto@case-research.it

G. Succi
Innopolis University, Innopolis, Russia
e-mail: giancarlo@giancarlosucci.com

© Springer International Publishing Switzerland 2016
P. Ciancarini et al. (eds.), *Proceedings of 4th International Conference
in Software Engineering for Defence Applications*, Advances in Intelligent
Systems and Computing 422, DOI 10.1007/978-3-319-27896-4_25

1   Introduction

The prediction of the fate of a requirement exhibits a strong impact on the use of the resources associated with it. During the lifetime of a software system, usually requirements are frequently modified or even completely deleted, due to factors such as change of business needs, increase of the complexity of the system itself, changes in mutual dependencies, or disability to deliver the service [1, 2]. Starting the implementation of a requirement that does not get inserted in the final product implies that we not only waste the resources associated with it, but also we create code and other artifacts that may be difficult to identify and delete once the decision of not to proceed with such requirement has been made, if it will ever been made explicitly instead of simply letting the requirement to fall in the oblivion. Moreover, an additional requirement creates an increased complexity in managing the system underdevelopment.

Therefore, it would be of immense benefit to predict the fate of such requirement as early as possible. In this way, also the management of the processes would be more cost-effective and deterministic [3]. Still, such prediction does not exist and has not been subject yet to a significant investigation.

Empirical software engineering is an established discipline that employs when needed predictive modeling to make sense of current data and to capture future trends; examples of use are predictions of system failures, requests for services [3], estimation of cost, and effort to deliver [4, 5]. With respect to requirement modeling, empirical software engineering has so far put most effort in life cycle modeling [6–8] and volatility prediction [9, 10]. In our view, the prediction of the fate of requirements and of the associated costs has not received an adequate attention; in this work, we propose this novel viewpoint that centers its attention on this neglected area of study.

In this study, we propose an approach to build such a prediction in a domain in which it is of particular benefit—the embedded systems domain, where typically the cost of errors is higher due to its direct impact on hardware. To determine whether a requirement will fail, we consider simply the history of the operations it underwent, treating it as a black box. We use logistic regression to discriminate the failure.

In this work, we use logistic regression to learn from historical sequences of data, each sequence with a final label of success or a failure depending on its final evolution state in an industrial software system development process. The model is experimented on a real data coming from 10 years of requirements properly evaluated in an Italian company operating in an embedded domain. The experimental data feature more than 6000 requirements and 80,000 logs of their evolution.

As it is shown further, this results in a statistically significant model that is able to detect failures and successes with a satisfactory precision. This analysis opens the path for a future better understanding of the requirement lifecycle dynamics, the identification of the typical patterns that show a future failure, and a broader exploration of classifying techniques for data of software engineering processes.

Moreover, keeping in mind the limitations of the model we build, the next step would be also a consideration of the trade-off between specificity and out-of-sample generalization.

The paper is structured as follows: In Sect. 2, we present the state of the art, and in Sect. 3, the methodology followed in this early study is detailed, highlighting the prediction method chosen and the way its performance is assessed. In Sect. 4, it is presented the case study. The paper continues further with Sect. 5, where the limitations of this work are outlined and the future work is presented. The paper is concluded (Sect. 5) by summarizing the results of this work and its importance.

2 State of Art on Empirical Studies on Requirements

Requirements are an important part of the software development process; some of their characteristics are studied broadly, and the results of such studies are widely present in the literature. However, as mentioned in the introduction, there is more to explore especially in terms of empirical studies of requirements evolution [11]: In this context, the existing research is focused on two main directions. The first is modeling the requirements evolution with the aim of understanding its dynamics early on their specification [6] or during the life cycle [8, 12], considering important and also developing tools to capture this evolution through logs of different stages of development process [13].

The second direction has for long focused on the requirements volatility. There are studies identifying some main reasons that can take a requirement not to succeed or to change frequently, business needs, requirement complexity and dependency, environment of development, etc [1, 2]. To the volatility issue, a few studies approach it, as a reflection of the sociotechnical changes in the world and in the industry. For instance, Felici considers changes not an error to deal with, but an unavoidable part of software process development that reinforces the requirement itself and saves the project from failing, by adapting continuously. In this case except the cost accompanying the requirements volatility, the authors try to associate also the benefit driven from the changes [14–16]. There are also several studies that focus importance to how changes propagate [17], and how they can be related to an increase in defects present and in the overall project cost [18].

With respect to learning from historical data, there are empirical software engineering studies attempting to build predictive models on requirements, some using simple regression models for identifying which are the future volatile trends and which requirements are more prone to change [9, 10], and other using different alternatives like building a rule-based model for prediction [7]. Describing changes and stability points through scenario analysis is also another proposed solution [19].

In any case, what is relevant to stress with reference to our work is that often the deleted requirements are not taken in consideration at all in these analysis, or the deletion process itself is considered simply as one of the changes that may happen [14, 10]. Thus, despite being a fairly easy concept to grasp and challenging to work with, there is almost no literature referring to failing requirements or trying to model the requirement final state. We think that our unicity gives more value to this paper since it shift the attention toward a field explored so little.

3   Methodology

The main goal of this study is to answer to the research question: *Can we predict reliably the final status of a requirement of a software system analyzing its historical evolution through stages and the development environment? If so, how early on the life cycle of the requirement can we perform the prediction?* Our aim is to build a model to make this possible. We start by making a relevant assumption that requirements are independent of each other, meaning that the status of one requirement does not affect the status of other requirements. This assumption is quite stringent, and for certain perspective also unfeasible in the real world, still it is the basis of the follow-up analysis, and, as we will see, its violation does not affect the validity of our findings.

3.1   Prediction Method

Machine learning is usually the right technique when we do not have an analytical solution and we want to build an empirical one, based on the data and the observable patterns we have [20]. Inside this field, supervised learning is used when these data are labeled, meaning the dependent variable is categorical. This is also our case, since we are dealing with a dichotomy problem, a requirement being successful or failed; thus, the main choice is to use supervised learning algorithms in order to build a predictive model. Applying logistic regression algorithm, from the supervised learning pool, would produce not only classification of the requirement objects in a binary pass/fail, where pass stands for a success and fail for a never deployed requirement, but also the probability associated with each class.

Logistic regression is an easy, elegant, and yet broadly used and known to achieve good results in binary classification problems. Assuming that we have d predictors and n observations, the predictors are formed on a basis of X and Θ with X being the matrix of observations and Θ the matrix of the weights:

$$\text{for } X = \begin{bmatrix} x_{11} & \cdots & x_{1d} \\ \cdots & & \\ x_{n1} & & \end{bmatrix} \quad \theta = \begin{bmatrix} 1 \\ d \end{bmatrix} \quad \text{the LR model is} \quad \text{logit}(X\theta) = \frac{1}{1+e^{-X\theta}}$$

In order to estimate Θ, the algorithm minimizes the uncertainty (out performance index), by first assuming an independence between the observations. The result of the iterations is the discriminant that divides the input space into pass and fail regions. Thus, per any new requirement data point, once learned the model and tuned with the right threshold, we would be able to give a statistically significant probability value of it being a future pass or fail.

3.2   Evaluation of Performance of Predicting Models

To assess the performance of the classifier, we use the confusion matrix. The confusion matrix values allow us to calculate the predicting ability and error made in terms of type I and type II error. Clearly, since logistic regression delivers a probability of belonging to a certain class, as we discussed in the section above, in order to determine the label, a threshold is used. As the threshold changes, also the ability of classifiers to capture correctly passes and failures is channeled (inclined/biased). We use the ROC curve analysis, through maximizing the combination of sensitivity–specificity of our model, and we obtain the best threshold value and tune the model accordingly. Thus, the metrics we will use to measure performance are as follows: sensitivity, specificity, and the area under the curve. Sensitivity shows the ability of the model to capture correctly fail requirements in the dataset, whereas the specificity shows the ability to capture passes. The area under the curve is calculated by plotting sensitivity versus specificity for threshold value from 0 to 1, whereas the formulas to calculate these two are below:

Confusion matrix	Predict fail	Predict pass
Real fail	True positive	False negative
Real pass	False positive	True negative

$$Sensitivity = \frac{TruePositive}{TruePositive + FalseNegative}$$

$$Specificity = \frac{TrueNegative}{TrueNegative + FalsePositive}.$$

4   Case Study

4.1   Data Description and Processing

The data we use in this research come from an industrial software system in embedded domain. The data of the development process are formatted as a file of logs, presenting the history of 6608 requirements and 84,891 activities performed on them on a time period from March 2002 to September 2011. Every log is characterized by the following attributes summarized in the table.

| Date and time in which the requirement was firstly created in the log system |
| Version of the software product in which the requirement was first created |
| Developer that performed this specific action on the requirement object and entered it in the log system |
| The specific activity performed on the requirement |
| The identification number, unique per each requirement |

There are 13 different activities performed on a requirement throughout the development process, with the most significant numerically and interesting from the process point of view being:

| The requirement is created in the system |
| Creates a link between different requirement objects and keeps count of it |
| The requirement is modified |
| The requirement object is deleted |

The early identification of the deletion is basically what we are after in this study, since it means that the requirement was deleted from the system without ever being deployed, this is what we will call a "fail." Every other activity concluding the life cycle of the requirement present in database is considered a success and the requirement itself a "pass." At this point, it is clear that once processed, we get binomial labeling for the attributes characterizing the requirements life cycle inside this process.

From a first statistical analysis, we notice that there is a dominance of modification activities in the requirements labeled as "fail." The early attribute selection is performed based on domain expertise and also influenced by any possible pattern that we identified, as mentioned above. The processed final requirement records are the rows of a data frame whose columns are the potential features of the predictive model we aim to build, except the ID that is used to keep track of the specific object.

Each requirement record is the characterized by elements presented in the following table.

| The identification number |
| The version where it was first created; it will also ensure a timeline on the requirement objects |
| The last developer that worked with the requirement |
| The cumulative number of activities performed on the requirement; this is also an approximation of the number of modification a requirement goes through |
| The longevity of a requirement, calculated as the difference between the closing and the opening dates |
| The label either pass or fail |

4.2 Learning of the Model and Evaluation of Performance

We start with a slightly imbalanced time-ordered dataset of object profiled by four features that we will try to use in our model: version, number of transitions, lifetime, and developer. We allow the lack of balance in order to give an initial bias toward the pass class, considering that, however, a false pass costs less to the system than a false fail.

To learn the model, stepwise regression is performed on the data. The statistically significant features used by the model are as follows: version, number of transitions, and lifetime of the requirement.

The steps followed are as follows:

1. Dataset is divided into training and validation sets (75–25 %, respectively), preserving the initial class distribution present on the original data, using package "caret."
2. Logistic regression algorithm is run (10-k cross validation, repeated 3 times) on training data to build the model.
3. ROC curve is plotted in order to find the best sensitivity–specificity values by using the criteria "most top left" on the ROC curve. We notice 79 % predictive ability of the model (AUC = 0.79 with 95 % CI) and the best threshold being 0.27.
4. Model is tested on the validation set, where for the best threshold value, the results achieved in terms of predictivity are as follows: sensitivity = 0.74 and specificity = 0.95.

Fig. 1 Performance of the developed model

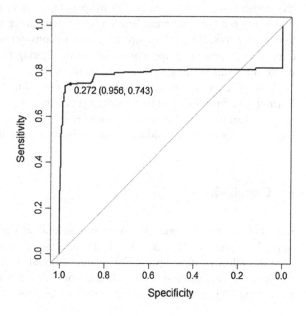

Thus, the model built out of all true failures can capture correctly 74 % of them and out of passes 95 % of them, with the results being under 95 % CI. Figure 1 shows the plot.

This classification model built and tested on the whole dataset of requirements shows clearly for a possibility of predicting pass and fail requirements.

5   Limitations and Future Work

This study delivers a novel idea that can move forward in several directions, especially by taking into account its limitations.

The first limitation is inherent with the single case study. This work has a very significant value given the size of its dataset—a very rare situation in software engineering and a quite unique situation in the case of requirements evolution; however, it is a single case, and as usual a replication would be strongly advisable.

The second limitation is that we have assumed the requirements independent, while they are not. This situation is similar to the one in which a parametric correlation is used, while clearly its requirements are not satisfied—still it has been experimentally verified that the predictive value still hold, as it holds in this specific case.

The third limitation is that we build a model having available the whole stories of the requirements, while it would be advisable to be able to predict early whether a requirement is going to fail. Therefore, the future studies should focus on trying to build predictive models using only initial parts of their story.

The fourth limitation is that we do not take into account that the information on requirements is collected manually by people, so it is subject to errors, and this study does not take into consideration how robust the requirements are with respect to human errors. So a follow-up study should consider whether the presence of an error, for instance in the form of a white noise altering the logs of the requirements, would alter significantly the outcome of the model.

A future area of study is then to apply more refined techniques from machine learning and from fuzzy logic to build a predictive system. The joint application of machine learning and fuzzy logic would resolve by itself the last three limitations mentioned above, also building a more robust and extensible system.

6   Conclusions

In this work, we have presented how we can analyze the evolution of requirements to predict those that will fail, thus building a model that could to decide where to drop the effort of developers. The model is based on logistic regression and has validated on data coming from an Italian company in embedded domain, including more than 6000 requirements spanning more than 10 years of development.

The results of this early study are satisfactory and pave the way for follow-up work, especially in the direction to make the model predictive, more robust, on employing more refined analysis techniques, like those coming from data mining and fuzzy logic.

References

1. Harker SDP, Eason KD, Dobson JE (1993) The change and evolution of requirements as a challenge to the practice of software engineering. In Proceedings of 1st ICRE, pp 266–272
2. McGee S, Greer D (2012) Towards an understanding of the causes and effects of software requirements change: two case studies. Requir Eng 17(2):133–155
3. Succi G, Pedrycz W, Stefanovic M, Russo B (2003) An investigation on the occurrence of service requests in commercial software applications. Empirical Softw Eng 8(2):197–215
4. Boehm B, Abts Ch, Chulani S (2000) Software development cost estimation approaches—a survey. Ann Softw Eng 10:177–205
5. Pendharkar P, Subramanian G, Rodger J (2005) A probabilistic model for predicting software development effort. IEEE Trans Softw Eng 31(7)
6. Yu ESK (1997) Towards modelling and reasoning support for early-phase requirements engineering. In: Proceedings of 3rd IEEE international symposium on requirements engineering, pp 226–235, 6–10 January 1997
7. Le Minh S, Massacci F (2011) Dealing with known unknowns: towards a game-theoretic foundation for software requirement evolution. In Proceedings of 23rd international conference: advanced information systems engineering (CAiSE 2011), London, UK, 20–24 June 2011
8. Nurmuliani N, Zowghi D, Fowell S (2004) Analysis of requirements volatility during software development life cycle. In Proceedings of the ASWEC, pp 28–37
9. Loconsole A, Borstler J (2007) Construction and validation of prediction models for number of changes to requirements. Technical report, UMINF 07, 03 February 2007
10. Shi L, Wang Q, Li M (2013) Learning from evolution history to predict future requirement changes. In: Proceedings of RE 2013, pp. 135–144
11. Ernst N, Mylopoulos J, Wang Y (2009) Requirements evolution and what (research) to do about it. Lect Notes in Bus Inf Process 14:186–214
12. Russo A, Rodrigues O, d'Avila Garcez A (2004) Reasoning about requirements evolution using clustered belief revision. Lect Notes Comput Sci 3171:41–51
13. Saito S, Iimura Y, Takahashi K, Massey A, Anton A (2014) Tracking requirements evolution by using issue tickets: a case study of a document management and approval system. In Proceedings of 36th international conference on software engineering, pp 245–254
14. Anderson S, Felici M (2000) Controlling requirements evolution: An avionics case study. In: Proceedings of 19th SAFECOMP, pp 361–370
15. Anderson S, Felici M (2001) Requirements evolution from process to product oriented management. In Proceedings of 3rd PROFES, pp 27–41
16. Anderson S, Felici M (2002) Quantitative aspects of requirements evolution. 26th annual international computer software and application conference, COMPSAC 2002, pp 27–32, 26–29 August 2002
17. Clarkson J, Simons C, Eckert C (2001) Predicting change propagation in complex design. In: Proceedings of DETC'01, ASME 2001 design engineering technical conferences and computers and information in engineering conference, Pittsburgh, Pennsylvania, 9–12 September 2001

18. Javed T, Maqsood M, Durrani QS (2004) A study to investigate the impact of requirements instability on software defects. ACM SIGSOFT Softw Eng Notes 29(3):1–7
19. Bush D, Finkelstein A (2003) Requirements stability assessment using scenarios. In Proceedings of 11th ICRE, pp 23–32
20. Yaser SAM, Malik MI, Hsuan-Tien L (2012) Learning from data. Available online at URL: http://www.amlbook.com/support.html

Capturing User Needs for Agile Software Development

Stella Gazzerro, Roberto Marsura, Angelo Messina and Stefano Rizzo

Abstract Capturing the "user requirement" has always been the most critical phase in systems engineering. The problem becomes even more complicated in software engineering where uncertainty of the functions to implement and volatility of the user needs are the rule. In the past, this initial difficulty of the software production phase was tackled by the requirement engineers using formal methods to structure the dialogue with the user. The solutions offered by this discipline are controversial. Agile methods put once again the human factor at the center of the user needs capture phase, using the intrinsic nonlinearity of the human mind to translate the user stories into software development. Although this practice has a positive outcome, the built-in ambiguity of the natural language is a consistent limit. An attempt to structure this dialogue without disturbing the "Agile" approach is presented, and some considerations on the possibility of using semantic tools are made.

1 Introduction

The topic of the user needs capture and elaboration is tackled in the very challenging frame of a mission-critical C4I [1] "Agile" software development in the military domain. Commencing from the legacy software production cycle, some brief

Stella Gazzerro: Profesia, Angelo Messina: Italian Army General Staff.

R. Marsura · A. Messina (✉)
ReS Srl, Rome, Italy
e-mail: angelo.messina@esercito.difesa.it

S. Gazzerro · S. Rizzo
Emerasoft, Torino, Italy

© Springer International Publishing Switzerland 2016   307
P. Ciancarini et al. (eds.), *Proceedings of 4th International Conference
in Software Engineering for Defence Applications*, Advances in Intelligent
Systems and Computing 422, DOI 10.1007/978-3-319-27896-4_26

considerations on the problems connected with the implementation of the Agile methodology in the area of mission-critical software production are presented.

Focus is then put on the initial phase of the Agile production cycle that replaces the old-style user requirements gathering with the "user stories" collection. If on the one hand the use of the natural language gives the users more freedom and confidence in the software development activity, on the other hand it is introducing in the production cycle the uncertainty connected to language itself. It will be demonstrated that this uncertainty is still preferable to the introduction of any translation layer between the user needs and the developers.

An analysis of more than six hundred user stories has been conducted using both "human-assisted" and semantic tool-assisted. The major findings are briefly discussed. Here, conclusions and the way ahead are illustrated.

2   The Legacy Software Production Cycle

Comparing "waterfall" and "Agile" on the same budget allocation perspective, it is clear that the return on investments is radically different. In the first case, it might not be possible to see any code at all. In the second, it may conclude with a partial delivery, but carefully working out the product backlog, some initial functionalities could be delivered. The second major difference is connected to the requirements verification. In the traditional approach, the first real check of the product against the user requirement is performed very late in the process, whereas in Agile it happens very soon, even after a few weeks into the production process. Another element in favor of Agile is the reduced projects' failure rate as reported in the literature. Considering for reference the experience accumulated by the Italian Army in the development of the Command and Control software LC2EVO [2], a significant improvement in the effectiveness has been observed. Although the software cost reduction may be indicated as the most relevant factor for improvement, the Army stakeholders seem to appreciate the customer satisfaction increasing even more. This is a peculiarity of the evolution of the C2 [3, 4] function in the last years, which is characterized by a relevant instability of the user requirement often defined by the Army developers as the "volatile user requirement" problem. In the "old-fashion" production cycles such as the MIL STD 2167A [5], "waterfall," or similar, the time between the specification of the user requirement and the first result of the production cycle could be measured in years. In such a long period, the Army's approach to operations could change several times as well as the key features of the strategic view needed by the decision makers. The product designed on specification documents which are some years old cannot match the current user needs in this way an "Agile" software production environment is mandatory for a rapidly changing scenario.

For all the exposed reasons, the Army has decided to try new ways of implementing the software development life cycle. The choice initially fell on Agile

Scrum [6] which seemed suitable for a rapid implementation at the Army facilities. As a matter of fact, the acquisition of the "Agile" culture and the implementation of Scrum were not so easy as forecasted.

3 Adapting "Agile" to "Mission-Critical" SW Production

The commercial versions of the "Agile" methodology implementation (i.e., Scrum, Kamban, XP) [4] cannot be implemented as they are for the production of "mission-critical" software. The main reason is the need to guarantee the quality and security which has been embedded in the product since the very first line of code was written. The use of quick delivery cycles, of only three–four weeks, can easily result in setting the major development focus on customer satisfaction only giving less and less attention to the quality and security of the written code. This is a typical "Android apps" approach where the first law is: "do it quickly and expose the product on the market, and the bugs will be taken care off later." This is fully understandable for a production cycle that has the "must" to reach the potential customer before the competitors, but it cannot be accepted for a software product where reliability and security are extremely valuable for the users even if these embedded qualities do not show up evidently to the eyes of the customer.

4 User Stories Instead of Requirements

The substitution of the traditional "formal requirements document" with a living and changing nondocument can be very traumatic. The first two major problems strictly deriving from the freedom of speech given to the stakeholder are the use of the language (in this case Italian) and the semantic uncertainty that goes along with natural language [7].

Use as reference the basic structure of the user stories: "as ⟨user type⟩ I have to be able to ⟨action or capability⟩ in order to ⟨effect or end state⟩."

The idea that a complex set of operational functions can be explained in terms of a finite set of simple statements like the one at the base of the "Agile Scrum" methodology is, of course, a gross approximation.

As later reported in this document, the communication we are describing happens between two different human agents who are freely using their natural language; therefore, all the limitations and ambiguity connected to "human" communication apply.

The following text extracted from St. Augustin "Catechizandis rudibus" *the first cathechetical instruction* clearly shows how the problem of mind content representation through language was well known in 370 AC.

> **CHAPTER 2**
>
> *A disappointing experience: the inadequacy of language to express thought. But let the catechist take courage.*
>
> I am sorely disappointed that my tongue has not been able to answer the demands of my mind. For I desire my hearer to understand all that I understand; and I feel that I am not speaking in such a manner as to effect that.

4.1 Use of Natural Language (Italian)

The first problem mentioned above implies a simplified logic analysis (in Italian: *analisi del periodo*) of the user story text to identify the full structure of the sentence and the relationship between the principal and secondary statements (in Italian: *periodo principale e secondario*). In a pure conversational environment, the mentioned structure can easily degenerate in a set of multiply linked secondary statements which are capable of strongly affecting the effectiveness of the user stories collection phase.

In our study, the subject (implicit "I" first person) is identified to build the Users class which includes all the possible instances of "user." The capability or ability defined by the verb is then recognized to build the "Functionality" class. The third part of the sentence contains the effect or end state of an object as a result of the ability implementation. This is collected in the "Effect" class.

As it is reported in the attachment (see excel SS) not all the user stories analyzed can be reduced to the basic frame of the theory, but nevertheless a gross categorization in classes has been made.

The decomposition of the user stories has made it possible to evaluate them using a simple scorecard:

#	Criterion	Score (range)
1	Clearly identified subject	1–5
2	Syntactic quality	1–5
3	Independence	1–3
4	Completeness	1–4
5	Value (effects or end state)	1–5
6	Flexibility	1–3
7	Need of external definitions	3 (no need) to 0

The study has considered 3 of the 7 teams currently working at the Army General Staff on the LC2EVO software project. A total number of 600 user stories have been analyzed, decomposed, and evaluated. The stories are in Italian language, and the syntactical analysis has been made in full accordance with the standard Italian language rules (www.grammaticaitaliana.eu/analisi_logica.html). The translation given for the examples reported in this article is for demonstration use only since a coherent professional translation of the stories would be necessary to run a similar analysis with the rule of the English language. It is highly recommended to keep the user stories elaboration process as straightforward as possible avoiding any further translation into other natural languages or other metalanguages before a full understanding of the "true meaning" intended by the user with the story is reached.

As user	I want to be able to	In order to (business value)	Done definition
Of FAS Txxxxxx User story. #4 Team 6	Prepare in "assisted mode" all the relevant documents necessary to prepare the Joint Fires attachment of the Operation ORDER	Reduce the possibility of human errors and reuse all the geometric elements already available to the operational scenario (coordinates, deployment zones, coordination measures, and positions of the various elements on the battlefield)	The user will be able to define in assisted mode the content of attachment H of STANAG xx99

Criterion #1 defines the clear identification of the user (subject) in the story, on a scale 1–5 being five the maximum value when the subject is well defined and already known to the specific software development. The criterion contains an implicit "reward" for stories which are easy to connect to the segment of the product already delivered (Table 1).
Criterion #2 Syntactic quality is a pure language evaluation criterion linked to the natural language ambiguity used to add value to the stories that have a simple and fully understandable (unambiguous) structure.
Criterion #3 Independence is a Scrum Agile derivative criterion used to evaluate the self-consistence of a story without the need of elements belonging to other US.
Criterion #4 Completeness evaluates the possibility of defining the software development task in an exhaustive manner on the base of the single story.
Criterion #5 Value (effects or end state) is an evaluation of the content of the story in terms of business values and/or end state of the software functions implemented

Table 1 Summary of US evaluation on the 7 criteria

US #	Team#	C1	C2	C3	C4	C5	C6	C7	Total score
4	6	5	3	2	4	4	1	1	20

or modified as an effect of the story elaboration. The basic question answered by this criterion is: "Is the user able to recognize the (operational/functional) value implemented after the story related tasks accomplishment? Is this reflected in an agreed "done" definition?"

Criterion #6 Flexibility is used to assess the risk of taking the story on board. A flexible story should not generate too many constraints for the software development. This is typical of many "nonfunctional" stories/requirements such as the selection of a graphic environment or a database solution.

Criterion #7 Need of external definitions is a "reward" mechanism for self-describing stories which contain all the data necessary to proceed to the task definition. The neutral value (0) examples are the stories which are complete and logically exhaustive (do not imply further developments) but need long and articulated attachments as STANAG or standards.

The analysis whose results are only partially reported in table (…) shows that the chosen criteria are able to capture some of the user stories collection process characteristics. Further refinement of the scorecard is needed, and a possible automation of the process is envisioned.

As lesson identified on the application of natural language to the user needs capturing phase in our specific experience, the first consideration that can be made is concerning the very nature of the "Defense Command and Control" domain. It is clear that the basic structure of the "Scrum" user story is not capable of capturing the majority of the user needs. This is mostly due to the complexity of the C2 functions. The frame "as a $\langle ... \rangle$ I have to be able to $\langle ... \rangle$ in order to $\langle ... \rangle$" cannot capture the entire essence of a story as in the example presented in the picture above.

The story presented in the example is one with a medium–high quality level taking advantage of one year of improvement in the process. Nevertheless, in presenting the user side of this artifact, it is quite evident that the apparently simple statements are hiding a completely different view of the same world between the operational user and the developer. While the "user" of the story is clearly identified (score 5 on C1), with a single definition: "assisted mode," the user is summoning some kind of automated process used to minimize the possibility of mistakes, but he/she is not specifying the real functions to be automated. This is clearly reflected in the "Done definition" which is almost undistinguishable from statement 1. This language ambiguity is reflected in the score (3 on C2).

The story is not completely independent either, since it is clearly referring to geometric entities developed (or to be developed?) in other US (score 2 on C3).

Given the limitation explained above, the story is fairly complete toward a task allocation by the developers (score 4 on C4).

The business value of the story is very high in the perception of the user who expects some kind of "error-free tool" (score 4 on C5).

Flexibility of the story appears to be very low, since none of the suggested functionalities can be negotiated. The developers will also have to acquire confidence on the quoted STANAG on their own, and this is reflected in the low scores on the last two criteria (C6 and C7).

The analysis clearly shows a growth in the story quality as team progresses into Sprints. This is due to various factors: The story collection process improves as the knowledge of the problems increases; the stakeholder gets more confidence in the "Agile" interaction; teams and stakeholders develop a "common language," which is basically a subset of the natural language (dialect) where word semantic is "tuned" to the specific domain (some word meaning is actually redefined).

4.2 Semantic Uncertainty

Even though a discussion on the conceptual semantic implications of dealing with "natural language" is not in the scope of this paper, some considerations directly connected to this area of linguistics have to be taken into account. There is no doubt that the communication in natural language between members of different "worlds" (the operational stakeholder's and the software developer's one) makes it necessary to understand at least some of the criticality involved in the representation of the reality they give. Our analysis shows that there is a form of reconciliation between the two different visions of reality, mainly driven by the brief delivery time between software increments typical of the "Agile" methodology, but this is highly dependent on the people involved.

To give an example of the difference between the semantic meaning of a nominal identical concept, we should consider the notion of "battle space geometry." Using a simple user story template, an operational user would say:

as a command post officer I want to be able to represent the forward line of my sector on a map.

The user has in mind a "line" whose geometrical elements (waypoints, start, finish, and type) are decided by a superior command post that is given to him as part of a formatted order packet which he expects to appear on the map by a single click of his mouse and to be updated as the tactical situation changes.

The developer's first comprehension will be "drawing" tools able to produce a line by simply calling some graphic libraries. The software engineer focus will be on how to implement the graphical feature writing the least possible quantity of new code. This is just an example but qualifies the differences between the two worlds very well.

For the user, the image on the video is just a representation of his/her real world and has in mind a real portion of land where the operation is taking place, whereas for the developer the same object is the result of a series of software–hardware interactions (his/her world) showing a map on a video where he/she has to provide a designing tool.

As trivial as it may seem, these differences are at the root of the software requirement specification problem that in the past has been tackled by freezing the "requirement text" and operating a certain number of translations into formal representations without reconciliation of the two different "world representations." It has not come as a surprise that so many software projects have failed.

Some concepts developed in conceptual semantics explain how the representation of the world expressed by natural language is the result of mediation between the speaker's perception of the world and the current setup of his/her own mind. That poses the question on what we really do communicate with natural language. Again, further discussion of these well-known issues is out of the scope of this paper but nevertheless must be taken into account if an improvement of the user stories collection process has to be investigated.

The peculiar nature of the software product introduces some simplification in the above-mentioned linguistic problem and gives some opportunities as well. Finally, software is just another language used to describe a reality which is very close to the real world. The Agile methodology with its frequent incremental delivery is providing step after step a description of the real world that is shared by operator coming from different approaches.

In terms of knowledge, the reference ontology could easily be the same both for users and for developers. Finding a way for working on the ontology in the initial production process would improve the effectiveness of the Agile methodology. The user would then have a synthetic representation of his/her needs easy to maintain and sharable with software engineers.

5 Automated Link Discovery Between User-Relevant Entities with Semantic Tools

The previously presented "human-assisted analysis" is a coherent benchmark for the application of the state-of-the-art tools for information management and knowledge management.

In the last few years, two relevant market trends, Big Data and Semantic Web, have pervaded the global research community and generated high expectations on the market [n]. One of the most interesting promises that these trends are currently nurturing is the ability to discover hidden relationships among data that could not be perceived before by the human brain for the simple reason that data were not available and, when available, too big for the human comprehension. Besides being crucial for intelligence, and strategic and tactical purposes, there is a wide consensus [n] that could be worth verifying the power of new emerging semantic analysis algorithms to extract new information from the broad knowledge that lies in user stories and project specification data.

The research project that started a few months ago spreads from the idea that a semantic analysis of project entities could represent a new perspective on project information and could reveal hidden relationships that could feed some key practices such as:

- knowledge retrieval,
- information reuse,
- data homogenization,
- project management intelligence.

Therefore, the approach is to support automated information retrieval and "hidden" link discovery aiming at the automated creation of a meaningful semantic ontology.

The initial results of our experiments are very encouraging. For each user story in the system, it has been possible to find semantically similar items:

Selected

3d12adcd-6358-4293-af46-74f3402fc918 – Accesso dedicato a Gestione Infrastrutture – Elenco User Story-2.docx
L'utente deve accedere in modo diretto a Gestione Infrastrutture e non attraverso il Visualizzatore.   L'obiettivo di questa attività è disaccoppiare le applicazioni.

Similar

●● 269
0e7ea519-9224-491e-9509-33dfa417207d – Visualizzatore Desktop – Elenco User Story-2.docx
Come utente, devo avere a disposizione un visualizzatore desktop Visualizzatore Desktop Come utente, devo avere a disposizione un visualizzatore desktop

●● 236
1b75ac24-10e7-432d-b044-27406a29dc04 – creazione sezioni – FAS – Elenco User Story-2.docx
Devo poter visualizzare una sezione del mio portale che mi permetta di accedere alle FAS. creazione sezioni – FAS Devo poter visualizzare una sezione del mio portale che mi permetta di accedere alle FAS.

●● 198
53cf3a21-518c-4c8d-b537-ea98da1ed02d – dati bussola – Elenco User Story-2.docx
Come utente, devo poter visualizzare il dato relativo alla bussola sul

Similarities were then grouped in an automatically generated graph of semantic proximities that represent new relevant relationships between items (nodes) in the source information.

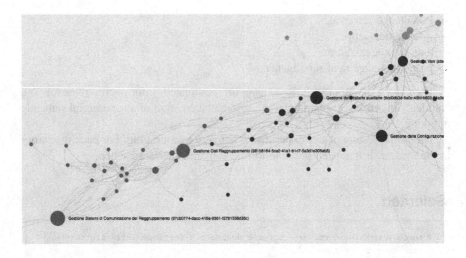

These new relevant relationships are proving to be true and valuable, revealing new interesting knowledge from source data.

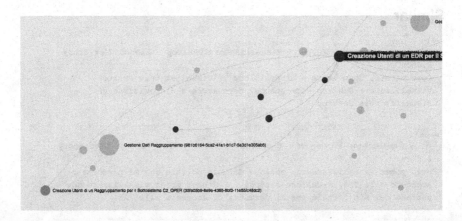

So, from an undefined mass of data, we can generate a new organization of relevant information linked together.

Another key benefit of the semantic analysis is the generation of a semantic index of detected tags:

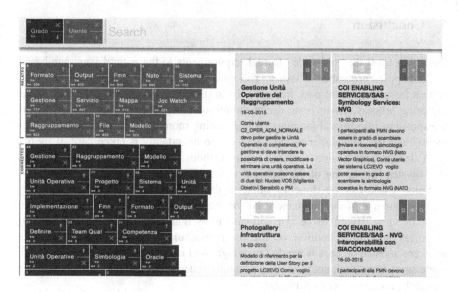

It is also possible to navigate the information by adding and removing the semantic tags automatically detected:

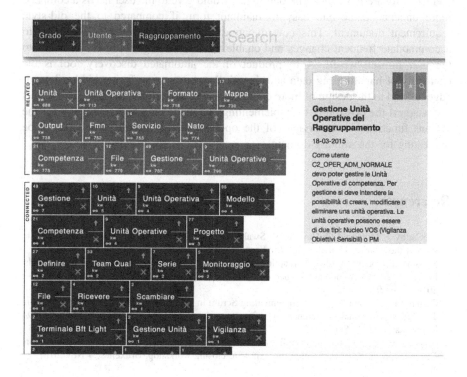

6 Conclusion

Switching from the "old-style" formal requirements to the natural language-based methods poses a number of issues. The intrinsic uncertainty of the language, well known from the ancient past, has never been solved and is affecting the user needs capturing process. There is no doubt though that any layer of translation such as transposition into formal language is introducing more distance between the "user world" and the "developer world." Part of the uncertainty is taken care of by the short delivery production cycle of Agile methods which is performing a kind of step-by-step fine-tuning of the different views of reality.

Some work can be done to mitigate the linguistic/semantic uncertainty by studying the use of the language used for the user stories from a purely linguistic point of view, trying to design software tools that could limit the structure of sentences and the repetition of the used vocabulary items. It is clear that a software-assisted process is needed to analyze and keep track of the user stories life cycle since from such analysis, it is possible to derive significant measures on the team effectiveness and performance.

The automated discovery techniques can be successfully used to discover a link between the user stories content and the delivered software product and may bring one to the construction of specific ontologies connecting the users to the implemented software functions. The ontologies would give to the user needs a compact, more understandable and easy to maintain form if compared to the old-style requirement document. This new dynamic kind of requirement will be able to accommodate frequent changes and enable a better understanding of the needs to the user himself/herself. A by-product of the automated discovery tool is the capability of handling big data and though the ability of surfacing hidden knowledge, to lead to the design of more complete and meaningful ontologies. The results also suggest the possibility of implementing automatic ontology generation for the various functional area services of the operational C2 domain which would be interesting for the military domain.

References

1. McCarthy J, Morefield C, Gawron V, Swalm T, Davis R (1998) Information management to support the warrior. United States Air Force Scientific Advisory Board, ADA412936
2. Messina A, Cotugno F (2014) Adapting Scrum to the Italian army: methods and (open) tools. In: The 10th international conference on open source systems, San Jose, Costa Rica, May 6–9 2014. ISBN 978-3-642-55127-7
3. Cotugno F, Messina A (2014) Implementing Scrum in the army general staff environment. In: The 3rd international conference in software engineering for defence applications—*SEDA* Roma, Italy, 22–23 Sept 2014
4. Messina A (2014) Adopting Agile methodology in mission critical software production. Consultation on cloud computing and software closing workshop, Bruxelles, 4 Nov 2014

5. Changes from DOD-STD-2167A to MIL-STD-498 USA Department of Defence
6. Schwaber K (2004) Agile project management with Scrum. Microsoft Press, ISBN 9780735619937
7. Ray Jackendoff Foundations of language *Oxford University Press ISBN: 9780199264377*

A Course on Software Architecture for Defense Applications

Paolo Ciancarini, Stefano Russo and Vincenzo Sabbatino

Abstract What is the role of a software architecture when building a large software system, for instance a command and control system for defense purposes? How it is created? How it is managed? In which way can we form software architects for this kind of systems? We describe an experience of designing and teaching several editions of a course on software architecture in the context of a large system integrator of defense mission-critical systems—ranging from air traffic control to vessel traffic control systems—namely SELEX Sistemi Integrati, a company of the Finmeccanica group. In the last few years, the company has been engaged in a comprehensive restructuring initiative for valorizing existing software assets and products and enhancing their productivity for software development. The course was intended as a baseline introduction to a School of Complex Software Systems for the many software engineers working in the company.

1 Introduction

When developing a large complex software-intensive system, a central concept is that of software architecture.

Once the requirements specification phase is complete, design can start. The main goal of the design is to describe how a system will have to be built in order to meet the specified requirements. Other than functional requirements, the design

P. Ciancarini (✉)
Dipartimento di Informatica, University of Bologna, Bologna, Italy
e-mail: paolo.ciancarini@unibo.it

S. Russo
University of Napoli Federico II, Naples, Italy

V. Sabbatino
SELEX ES, Rome, Italy

© Springer International Publishing Switzerland 2016
P. Ciancarini et al. (eds.), *Proceedings of 4th International Conference in Software Engineering for Defence Applications*, Advances in Intelligent Systems and Computing 422, DOI 10.1007/978-3-319-27896-4_27

must meet the specified non-functional requirements, such as performance constraints, and constraints on dependability attributes, or power consumption constraints; moreover, these goals can be met at minimum manufacturing cost [2]. The first step of the design is to produce a high-level description of the system architecture. The architectural description is a plan for the structure of the system that specifies, at high-level, which components have to be included to carry out the desired tasks and their logical interconnection.

An initial architectural description is usually a diagram that does not distinguish hardware from software features, but rather focuses on the functionality required to each block in the diagram; in a successive phase, the initial block diagram is refined into two block diagrams, one for hardware and another for software [25]. These have to be designed to meet both functional and non-functional constraints. A notation for modeling both hardware and software architectural description is SysML [13]. In this phase, it is crucial to be able to estimate the non-functional properties of components in order to compare and evaluate different architectural solutions.

Making an architectural change in later phases would results in very high costs. The adopted models, formalisms, technologies, and tools in this phase should therefore provide the necessary means to carry out such estimations in easy and inexpensive way. Once the overall architecture has been outlined, the design efforts focus on the components of the architecture.

What is the role of a software architecture when building a large system, for instance a command and control system? How it is created? How it is man-aged? These questions are especially important for large engineering companies which produce families of software-intensive command and control systems [2]. A related question is how can one introduce and teach software architecture and asset reuse to engineers who have been designing systems for years without explicitly dealing with the concept.

The increase of software size and complexity raises critical challenges to engineering companies which build systems to be integrated into larger systems of systems. Production and management problems with software-intensive systems are strictly related to requirements engineering, software design, systems' families management, and their continuous testing integration and evolution.

A number of methods and solutions to these problems are based on the introduction of software architecting in the industrial practice. However, to become effective an architecture-centric development approach needs to be assimilated and put in everyday practice by the company personnel, who need architectural education and training.

In this paper, we describe our experience in designing and developing a course on software architecture for SELEX Sistemi Integrati (also known as SELEX SI, now Selex ES), an Italian company of the Finmeccanica group. The course has been taught in several editions from 2008 to 2015, and it is one of the courses set up in the framework of the public–private cooperation between Finmeccanica and major Italian universities organized in the CINI Consortium.

The paper has the following structure: Sect. 2 presents the company and the market context for which it develops products and systems. Section 3 discusses the requirements and rationale of the course. Section 4 describes the structure of the class and the activities which were performed. Section 5 summarizes the learning results. Finally, Sect. 6 discusses the results of this initiative.

2 The Context

The context of this experience is one of the most important Italian companies for system engineering: Finmeccanica, a large industrial corporate group operating worldwide in the aerospace, defense, and security sectors. It is an industry leader in the fields of helicopters and defense electronics. It employs more than 70,000 people, has revenues of 15 Billion Euros, and invests 1.8 Billion Euros (12 % of turnover) a year in R&D activities.

From Jan 2013, Selex Electronic Systems (in short Selex ES) is Finmeccanica's flagship company focusing on the design of systems. Selex ES inherits the activities of SELEX Elsag, SELEX Galileo, and SELEX SI to create a unified company. The new company operates worldwide with a workforce of approximately 17,900 employers, with main operations in Italy and the UK and a customer base spanning five continents. As the company delivers mission-critical systems in the above fields by utilizing its experience and proprietary sensors and hardware technologies—radar and electro-optic sensors, avionics, electronic warfare, communication, space instruments and sensors, UAV solutions—most of its engineers work on system integration projects and products. As software is becoming an increasing part of these systems, the need raised for updating and upgrading the employees' skills on architecting large-scale complex software systems.

SELEX SI (for the rest of this paper we will use this old company name because the course we describe was commissioned by that company) spends literally millions of hours in software development and integration inside its lines of products and systems. One of the main ways for minimizing development costs and maximizing the reuse of components consists of defining a reference software architecture for each product line [6]. In general, a product line reference architecture is a generic and somewhat abstract blueprint-type view of a system that includes the systems major components, the relationships among them, and the externally visible properties of those components. To build this kind of software architecture, domain solutions have been generalized and structured for the depiction of one architecture structure based on the harvesting of a set of patterns that have been observed in a number of successful implementations. A product line reference architecture is not usually designed for a highly specialized set of requirements. Rather, architects tend to use it as a starting point and specialize it for their own requirements.

We were required to define a course on software architecture strongly emphasizing the power of reuse product line reference architectures.

3   Designing the Course

The company started some years ago a learning program called SELEX Sistemi Integrati Academy. The Academy is an operational tool to enhance and to develop the corporate knowledge heritage and improve over time. The activities are developed by means of three schools: Radar and Systems School, System Engineering School, and School of Complex Software Systems, overall accounting for more than 100 courses.

The academic institution which has contributed to designing the School of Complex Software Systems is the Consortium of Italian Universities for Informatics (Consorzio Interuniversitario Nazionale per l'Informatica, CINI). The School of Complex Software Systems consists of 22 courses. A number of computer science and engineering professors of different CINI universities cooperated to its setup, teaching, and evaluation.

Software architecture was the first course of the School. The course was designed by a small committee including two university professors from CINI (the two first authors of this paper) and a number of people from SELEX SI, including engineering staff members (among them the third author) and the human resources department.

Software structure and architecture is becoming a classic topic for educating software engineers, as can be seen for instance in the latest version of the chapter on software design of the Body of Knowledge of Software Engineering. Also the teaching in an academic environment is quite mature, see for instance [14]. However, the personnel of large companies has a variety of experience and training histories and produce very specific architectures for specific markets. Thus, in an industrial context any teaching of software architecture has to take into account the specific requirements of the company managers.

The course committee mentioned above started from a set of learning requirements given by the company:

- the course should target generically all software developers active in SELEX SI. These are a few hundreds, with a variety of ages and cultural backgrounds. The basic assumptions were some proficiency in object-oriented programming in languages like C++ and concurrent programming in languages like Ada;
- the contents of the course should be customized for such SELEX SI personnel, meaning that the examples and the exercises should focus on real software systems developed by the company. This implied that the teachers should receive some information on the software of interest for the company. Moreover, a number of testimonials from actual projects would make short presentations showing how the main topics of the course were being applied in the company itself;

- the course, overall 35 h, should last one week (five days) or two half weeks, as chosen by the company according to the workload of the different periods.
- each class would include 20 people on average. Prerequisites for attending the class were the knowledge and practice of basic notions of object-oriented programming using a language like C++ or Java.
- all material should be written in English, as customary for all Schools in SELEX Academy.

The learning outcomes of the course on software architecture were defined as a review course on software systems architectures and related software engineering methods for designing complex software-intensive systems. The main idea was to establish a coherent baseline of software development methods, modeling notations, techniques for software asset reuse based on architectural styles and design patterns. More specifically, the learners were expected to obtain a basic knowledge of the principles governing software architecture, and the relationship between requirements, system architectures, and software components (Component-Based Software Engineering—CBSE).

The typical academic software architecture course consists of lectures in which concepts and theories are presented, along with a small modeling project whose aim is to give learners the opportunity to put this knowledge into practice. A text reference for this kind of course is [25]. Our own experience in teaching an academic course in software architecting is described in [5].

In an industrial context, these two activities, namely ex-cathedra lectures and toy project, are not adequate because learners have already some practical experience of large software projects and, more important, there is scarce time to present, compare, and put in practice a variety architectural theories. A reference for a course in an industrial context (Philips) is [16].

In our case, the syllabus was defined as an introduction to software architecture with some reference to software product lines, putting software architecting in a perspective of software reuse, and presenting the related technologies. More precisely, the course had to recall some advanced issues in software modeling and present the main ideas in architectural design for reuse. What follows is a short description of the contents of each lecture.

1. Introduction to software architecture and techniques for software reuse. This lecture introduces the main topic: How software-intensive systems benefit form focussing all the development efforts on the software architecture, aiming at reusing the existing assets for the various product lines architectures. The reference texts for this introduction were two: [7, 18].

2. Describing software architectures. This lecture had the goal to establish a methodological and notational baseline describing an iterative, architecture-centric process able to develop and exploit architectural descriptions for software systems using UML. Thus, its contents are the principles and the practice of modeling and documenting (software) systems with UML, aiming at the ability to describe, analyze, and evaluate a software system using a modeling notation. The reference text for this lecture was [11].

3. Reusable architectural design. This lecture has the goal to discuss the reuse of architectural design ideas: After a short recall of the elementary object-oriented design patterns (Larman's GRASP [15] and the more complex and well-known Gang of Four patterns [10]), the main architectural patterns are introduced and presented in a systematic way. An architectural style can be seen as providing the softwares high-level organization. A number of architectural styles have been identified that are especially important and critical for SELEX SI systems. The reference texts for this lecture were [12, 17].
4. Component and connector views in the context of pattern-oriented software architectures. This lecture has the goal to present some advanced examples of software architecture presenting their components and behaviors. The reference texts for this lecture were [3, 20].
5. Software systems architecture. This lecture has the goal to present some advanced issues of software architecture in the context of model-driven engineering. The main topics are the model-driven architecture and SysML, as used in SELEX SI. The reference text on MDA was [23]; on SysML [8, 9, 27].

4 Teaching the Course

We started giving this course in 2008, and until 2014 there have been six editions. For each edition, the same protocol has been followed. One month before the class week, the teaching slides prepared by the instructors were examined by personnel of the company, who could suggest some changes. After the modifications, the slides were put on line in a reserved site, and access was granted to the students. Thus, the people selected for the class could have a copy of the teaching material about one week in advance.

We had two variants of the timetable: class over one week and class over two half weeks. The one-week timetable is shown in Table 1.

When the course was split over two half weeks, each period had specific entrance and final tests.

Each edition leveraged upon the results of the preceding one, meaning that some adjustments were required to the course contents, usually triggered by some specific need presented by the company. Some suggestions came from the students'

Table 1 Structure of the class and version offered in one week

Time	Day 1	Day 2	Day 3	Day 4	Day 5
09–11	Test-in then lecture	Lecture	Lecture	Lecture	Lecture
11–13	Lecture	Lecture	Lecture	Lecture	Lecture
14–15	Testimonial	Testimonial	Testimonial	Testimonial	Testimonial
15–17	Hands-on	Hands-on	Hands-on	Hands-on	Lecture then test-out

comments, we gathered in class, other from people in the company covering the role of training manager. The suggestions impacted mainly the topic covered by the testimonials. In fact, these varied across the different editions of the course; the recurring topics they covered were as follows:

- The description of software architectures inside SELEX SI [19];
- Using UML for documenting SELEX SI systems;
- Typical architectural styles used inside SELEX SI;
- Experiences with MDA inside SELEX SI [1];
- Using SysML for SELEX SI systems.

All testimonial speakers were engineers of the company engaged in using the related technologies and methods in their projects.

Sometimes, some specific topics were raised and discussed during the lectures. For instance, the idea of architectural simulation [4, 24] was found very interesting and students required more information.

5 Results

The results were measured in two ways:

- Comparing the results of the entrance and final tests. A score sheet with 15 closed questions (each question had four possible answers) was given at the beginning of the first day and at the end of last day. The same set of questions and answers were used in both tests—entrance and final. In the first two editions (for which we have completed the analysis) the entrance test had a 41 % of correct answers, while the final test gave almost 80 % correct answers.
- Gathering the answers of the students to a satisfaction questionnaire. The satisfaction was measured on seven different parameters and averaged 82.7 on a scale 0–100. According to this figure, this course was one of the most appreciated in the School of Complex Software Systems.

6 Lessons Learnt and Conclusions

As mission-critical software systems become more complex, companies—in particular system integrators—feel an increasing need for improving their processes based on sound and effective techniques for software architectures definition and evaluation in specific domains [26]. This is what we observed at least at a nation-wide level in a number of public–private cooperations, which go even beyond the experience described in this paper.

When the field of software architectures emerged, it was argued it should be considered a discipline in its own [22]. After more than a decade [21], the corpus of scientific work in the field has developed consistently, but there is still a large gap between this and what is actually needed in industrial settings. Many of the achievements in the field have not matured enough; an example is architecture definition languages (ADLs) that we believe have not replaced—and are not likely anymore to replace—(extensions to) standard modeling languages. Some others are more mature, but they need to be tailored to the specific software processes, internal industrial standards, and practices.

Teaching software architectures in an industrial context requires to meet the company expectations in terms of mature knowledge and competences transferred to practitioners that they can subsequently turn into the practice of engineering complex systems. This is not easy to achieve, as architecting large-scale complex software systems—having tens of thousands of requirements and millions of lines of code—requires very high abstraction and modeling skills.

In our experience, successful teaching software architectures in an industrial context is not a unidirectional activity, but it benefits a lot from a strong cooperation between company stakeholders for the courses and teaching staff, and it included the following:

- Selecting and teaching the real foundations of software architecture, which have their roots in the very basic principles of software engineering;
- Selecting and teaching those architectural styles and patterns that can meet the real needs of practitioners in terms of design and reuse, with reference to their current target technologies; for instance, as modern middleware technologies are used in the design and development of large-scale software systems, this means defining and teaching proper guidelines that can help engineers to put architecture principles and patterns into their daily practice;
- Interleaving lectures and hands-on sessions with presentations by company testimonials, describing the practical problems, the current internal design standards, and the needs for more modern yet mature techniques;
- Tailoring the theory of software architecture to the specific domain and existing industrial software processes;
- Defining and teaching practical guidelines that engineers can apply reasonably easily to assess and evaluate software architectures based on project/product stakeholders' needs and predefined criteria.

We have also started a revision of our academic course in software architecture; a preliminary analytic comparison between the educational requirements of university students and company engineers is given in [5].

Acknowledgments The authors Ciancarini and Russo acknowledge the support of CINI. All the authors would like to personally thank Prof. P. Prinetto, director of the CINI School on Design of Complex Software Systems, and A. Mura, E. Barbi, A. Galasso, F. Marcoz, M. Scipioni, all from Finmeccanica companies, who in different stages and under different roles cooperated to the success of the various editions of the School.

References

1. Barbi E, Cantone G, Falessi D, Morciano F, Rizzuto M, Sabbatino V, Scarrone S (2012) A model-driven approach for configuring and deploying systems of systems. In: Proceedings of IEEE international conference on systems of systems engineering—SOSE. IEEE Computer Society Press, Genoa, Italy, pp 214–218
2. Bass L, Kazman R, Clements P (2012) Software architecture in practice, 3rd edn. Addison-Wesley, Reading
3. Buschmann F, Meunier R, Rohnert H, Sommerlad P, Stal M (1996) Pattern-oriented software architecture: a system of patterns, vol 1. Wiley, New York
4. Ciancarini P, Franzè F, Mascolo C (2000) Using a coordination language to specify and analyze systems containing mobile components. ACM Trans Softw Eng Methodol 9(2):167–198
5. Ciancarini P, Russo S (2014) Teaching software architecture in industrial and academic contexts: similarities and differences. In: Yu L (ed) Overcoming challenges in software engineering education: delivering non-technical knowledge and skills. Advances in higher education and professional development. IGI Global, pp 397–413
6. Clements P, Northrop L (2002) Software product lines. practices and patterns. Addison-Wesley, Reading
7. Clements P et al (2010) Documenting software architectures—views and beyond, 2nd edn. Addison-Wesley, Reading
8. Delligatti L (2014) SysML distilled. AddisonWesley/The OMG Press, UK
9. Friedenthal S (2008) A practical guide to SysML: the systems modeling language. Morgan Kaufmann/The OMG Press
10. Gamma E, Helm R, Johnson R, Vlissides J (1995) Design patterns. Addison-Wesley, Reading
11. Gomaa H (2005) Designing software product lines with UML. Addison-Wesley, Reading
12. Gorton I (2011) Essential software architecture. Springer, Berlin
13. Holt J, Perry S (2013) SysML for system engineering. IET
14. Lago P, van Vliet H (2005) Teaching a course on software architecture. In Proceedings of 18th IEEE conference on software engineering education and training (CSEET). IEEE Computer Society, pp 35–42
15. Larman C (2004) Applying UML and patterns: An introduction to object-oriented analysis and design and iterative development. Prentice-Hall, Englewood Cliffs
16. Muller G (2004) Experiences of teaching systems architecting. In: Proceedings of international council on systems engineering (INCOSE) symposium, pp 1–13
17. Qian K, Fu X, Tao L, Xu C (2006) Software architecture and design illuminated. Jones and Bartlett, Burlington
18. Rozanski N, Woods E (2012) Software systems architecture, 2nd edn. Addison-Wesley, Reading
19. Sabbatino V, Arecchi A, Lanciotti R, Leardi A, Tonni V (2012) Modelling the software architecture of large systems. In: Proceedings of IEEE international conference on systems of systems engineering—SOSE. IEEE Computer Society Press, Genoa, Italy, pp 1–7
20. Schmidt D, Stal M, Rohnert H, Buschmann F (2000) Pattern-oriented software architecture: patterns for concurrent and networked objects, vol 2. Wiley, New York
21. Shaw M, Clements P (2006) The golden age of software architecture. IEEE Softw 23(2):31–39
22. Shaw M, Garlan D (1996) Software architecture. perspectives on an emerging discipline. Prentice-Hall, Englewood Cliffs
23. Stahl T, Voelter M, Czarnecki K (2006) Model-driven software development: technology, engineering, management. Wiley, New York
24. Sterling L, Ciancarini P, Turnidge T (1996) On the animation of not executable specifications by prolog. Int J Softw Eng Knowl Eng 6(1):63–88
25. Taylor R, Medvidovic N, Dashofy E (2009) Software architecture. Foundations, theory, and practice. Wiley, New York

26. Valerio A, Succi G, Fenaroli M (1997) Domain analysis and framework-based software development. ACM SIGAPP Appl Comput Rev 5(2):4–15
27. Weilkiens T (2008) Systems engineering with SysML/UML: modeling, analysis, design. Morgan Kaufmann/The OMG Press